CONCISE DICTIONARY OF

MATERIALS SCIENCE

Structure and Characterization of Polycrystalline Materials

CONCISE DICTIONARY OF

MATERIALS SCIENCE

Structure and Characterization of Polycrystalline Materials

Vladimir Novikov

CRC Press
Taylor & Francis Group
Boca Raton London New York

CRC Press is an imprint of the
Taylor & Francis Group, an **informa** business

CRC Press
Taylor & Francis Group
6000 Broken Sound Parkway NW, Suite 300
Boca Raton, FL 33487-2742

First issued in paperback 2019

ISBN-13: 978-0-8493-0970-0 (hbk)
ISBN-13: 978-0-367-39580-3 (pbk)

This book contains information obtained from authentic and highly regarded sources. Reasonable efforts have been made to publish reliable data and information, but the author and publisher cannot assume responsibility for the validity of all materials or the consequences of their use. The authors and publishers have attempted to trace the copyright holders of all material reproduced in this publication and apologize to copyright holders if permission to publish in this form has not been obtained. If any copyright material has not been acknowledged please write and let us know so we may rectify in any future reprint.

Library of Congress Cataloging-in-Publication Data

Catalog record is available from the Library of Congress

Visit the Taylor & Francis Web site at
http://www.taylorandfrancis.com

and the CRC Press Web site at
http://www.crcpress.com

Preface

The present book is a new kind of a reference work for university and college students, as well as for self-educated readers with technical and nontechnical backgrounds.

It is common knowledge that mastering a speciality requires comprehensive reading of professional literature. My experience of more than 30 years of teaching students, supervising graduates and postgraduates, and consulting to industry, has enabled me to determine that one of the main problems in reading the specialized literature on materials science is a special kind of language barrier. In fact, the number of terms in the specialized literature is four to five times greater than in the textbooks and other educational aids. However, there appears to be no reference source available to help overcome this difficulty. Another problem, especially for self-educated readers, is establishing interrelations between the phenomena the terms denote. The available reference handbooks, containing mostly the definitions, offer little in the way of help. This book is designed to solve both problems: it bridges the terminological gap between the textbooks and professional literature while also affording the reader a coherent idea of structure formation and evolution.

Practically all the properties of various present-day materials are to a greater or lesser degree structure-dependent. This is true regarding traditional metallic materials in which, e.g., strength and plasticity are strongly affected by the dislocation density, grain size, number and size of second-phase particles, texture, etc. However, various physical properties of modern crystalline ceramics and semiconductors are also dependent on the presence of impurities and other lattice defects that largely affect the band structure. This is why the terminology on structure constitutes the bulk of the subject matter of this book. It contains about 1400 commonly used terms concerning the description of structure and its development, as well as the characterization of polycrystalline materials. Along with definitions, the majority of terms are accompanied by descriptions, explanations, and cross-references, thus providing a coherent picture of structure formation and evolution. The selection of terms for inclusion in this concise dictionary is based on the author's vast teaching and research experience. Emphasized are such principal topics as the lattice defects and their role in diffusion, plastic deformation, phase transitions, and distortions of band structure, as well as the granular structure, its formation and alterations in the course of phase transitions, plastic deformation, recrystallization, and grain growth. The terms specifying certain treatments and production procedures (heat and thermo-mechanical treatments, sintering, etc.) are presented in connection with their influence on structure. The terms relating to the modern investigation methods of crystal structure, microstructure, and local chemical composition are also included. In addition, certain crystallographic, thermodynamic, mechanical, and metallurgical terms used in structure description are found in the dictionary. The book also contains a list of acronyms popular in materials science.

The second part of this book, the English-German/German-English glossary, comprises the same entries as the concise dictionary. The glossary will meet the needs of a large number of readers working and studying in German-speaking countries, and the German-English section can be of value to English-speaking readers because many basic publications on materials science originally appeared in German. The combination of a bilingual glossary and a concise dictionary in one handy volume ensures quick access to the key terms and concepts.

Undergraduate, graduate, and postgraduate students studying materials science and engineering at universities and colleges, as well as members of training and refresher courses, will find this book invaluable. At the same time, it will be useful to research and technological personnel in metallurgical and metalworking industries. The book will offer great help to material-oriented physicists, researchers, and engineers developing crystalline materials for electronic applications. Moreover, chemists and engineers involved in microstructure research and the design of crystalline ceramics will discover a great deal of information usually lacking in textbooks, dictionaries, and reference books on ceramics.

The concise dictionary is also recommended as a reliable guide to nontechnical readers, such as managers, marketing and purchasing specialists, economists, insurance experts, and anyone else interested in materials science and engineering.

Vladimir Novikov
Hamburg, Germany

About the Author

Professor Vladimir Yu. Novikov studied materials science and technology at the Moscow Baumann Technical University and obtained his doctoral degree in materials science (1964) and his DSc degree in metal physics (1983) at the Moscow State Institute of Steel and Alloys (MISA). From 1962 to 1993, he delivered lecture courses on physical metallurgy, materials science, physical properties of metallic alloys, and special steels and alloys at MISA. Many of his former students are successfully working in industry and research in the U.S., Canada, England, Germany, and Russia. Over 100 scientific papers by Professor Novikov have been published in refereed journals. He is the author of two monographs: *Secondary Recrystallization* in Russian (1990) and *Grain Growth and Control of Microstructure and Texture in Polycrystalline Materials* in English (1996).

Using the Dictionary

Unless otherwise specified, the definitions of polysemantic terms given in the concise dictionary relate to materials science only.

The terms in the dictionary, in the list of acronyms, and in the glossary are arranged alphabetically on a letter-by-letter basis, ignoring spaces and hyphens. Combined terms are given without inversion. The entries are printed in boldface, and the definitions or equivalents are printed in regular type.

The cross-references in the dictionary are italicized. If there are synonyms, the most commonly used one is supplied with a description, followed by references to the others.

Asterisks affixed to some German terms in the glossary mean that the term is lacking in the German literature and the German equivalent given is a direct translation from the English.

List of Symbols

A energy of exchange interaction
A_f austenite finish temperature
A_N numerical aperture
A_s austenite start temperature
B_s bainite start temperature
C number of components (in Gibbs' phase rule)
D diffusion coefficient (diffusivity)
D_{gb} coefficient of grain-boundary diffusion
D_v coefficient of bulk diffusion
$\bar{D}$ mean grain size
D_M most probable grain size
D_{max} maximum grain size
D_{min} minimum grain size
E Young's modulus
F Helmholtz free energy
  number of degrees of freedom (in Gibbs' phase rule)
G linear growth rate
  shear modulus
  Gibbs' free energy
H enthalpy
  activation enthalpy
J diffusion flux
K constant of magnetic crystalline anisotropy
  bulk modulus
L designation of liquid phase
  length of dislocation segment in Frank-Read source
  Lorentz factor
M grain-boundary mobility
  Sachs or Taylor factor
M_f martensite finish temperature
M_s martensite start temperature
N ASTM grain size number
$\dot{N}$ nucleation rate
N_0 initial number of new phase nuclei
N_A Avogadro number
P number of phases (in Gibbs' phase rule)
  Larson–Miller parameter
Q released (absorbed) heat
  activation energy

Q^{-1} internal friction
R Rodrigues vector
R gas constant
   electrical resistance
S entropy
   long-range order parameter
T absolute temperature
T_0 equilibrium temperature
$T_0{}'$ equilibrium temperature of metastable phases
ΔT undercooling
T_C Curie point/temperature
T_g glass transition temperature
T_m melting point
T_N Néel point/temperature
U internal energy
a, b, c lattice constants
b Burgers vector
c_A atomic fraction of A atoms
c_v vacancy concentration
c_e equilibrium solubility limit
d mean size of second-phase particles
d_{hkl} distance between $\{hkl\}$ planes
d_{res} resolution limit
f volume fraction of second-phase particles
Δg driving (drag) force
h distance between dislocations in low-angle boundary
k Boltzmann constant
p polarization factor
r strain ratio
$\bar{r}$ Lankford coefficient
Δr coefficient of planar anisotropy
r_c dislocation core radius
r_{cr} critical size of nucleus
s true stress
t time
v_m migration rate
w dislocation width
Λ mean interparticle spacing
Θ disorientation angle
Θ_C Curie point/temperature
Θ_N Néel point/temperature
Σ CSL parameter
$\alpha, \beta...$ designations of phases in phase diagrams
α, β, γ axial angles
β_m metastable β-phase in Ti-based alloys
β_{tr} transformed β structure in Ti-based alloys

δ lattice misfit parameter
   logarithmic decrement
   grain boundary thickness
ε tensile strain
$\varphi_1, \Phi, \varphi_2$ Euler angles
γ true strain
   shear strain
γ_{gb} grain-boundary energy
γ_i energy of intrinsic stacking fault
γ_s free surface energy
λ wavelength
μ linear absorption coefficient
   chemical potential
ν Poisson's ratio
θ Bragg angle
ρ radius of curvature
   material density
ρ_d dislocation density
σ tensile (nominal) stress
   short-range order parameter
   interfacial energy
τ shear stress
τ_{cr} critical resolved shear stress
τ_{FR} shear stress initiating Frank–Read source
τ_p Peierls stress
τ_R relaxation time
τ_r resolved shear stress
τ_{th} theoretical strength at shear
ω Frank vector

List of Acronyms

AEM analytical electron microscope/microscopy
AES Auger-electron spectroscopy
AFM atomic force microscope/microscopy
APFIM atom probe field-ion microscope/microscopy
BCC body-centered cubic
CBED convergent beam electron diffraction
CCT continuous cooling transformation
CN coordination number
DIGM diffusion-induced grain boundary migration
DIR diffusion-induced recrystallization
DSC differential scanning calorimeter
DSC lattice displacement shift complete lattice
DTA differential thermal analysis
EBSP electron back scattered pattern
ECP electron channeling pattern
EDS/EDAX energy-dispersive spectrometry
EELS electron-energy loss spectrometry
EM electron microscopy
EPMA electron probe microanalysis
ESCA electron spectrometry for chemical analysis
FCC face-centered cubic
FIM field-ion microscope/microscopy
FWHM full width at half maximum
GP Guinier-Preston
HCP hexagonal close-packed
HIP hot isostatic pressing
HRTEM high-resolution transmission electron microscope/microscopy
HVEM high-voltage transmission electron microscope/microscopy
LEED low-energy electron diffraction
MFM magnetic force microscope/microscopy
ND normal direction (to the sheet surface)
ODS oxide-dispersion strengthened
OIM orientation imaging microscopy
PEEM photoelectron emission microscope/microscopy
PFZ precipitation-free zone
PLZT crystalline ceramic of composition $(Pb,La)(Zr,Ti)O_3$
PM powder metallurgy
PSN particle-stimulated nucleation
PZT crystalline ceramic of composition $Pb(Zr,Ti)O_3$

RD rolling direction
RDF radial distribution function
ReX recrystallization
SACP selected area channeling pattern
SAD/ESAD electron selected area diffraction
SAM scanning Auger-electron microscope/microscopy
SAP sintered aluminum powder
SEM scanning electron microscope/microscopy
SFE stacking-fault energy
SIBM strain-induced grain boundary migration
SIMS secondary ion mass spectroscopy
STEM scanning transmission electron microscope/microscopy
STM scanning tunneling microscope/microscopy
TEM transmission electron microscope/microscopy
TD transverse direction (in rolled sheets)
TRIP transformation-induced plasticity
TTT time-temperature-transformation
TZP tetragonal-[stabilized-]zirconia polycrystal
WDS wavelength-dispersive [x-ray] spectrometry
XPS x-ray photoelectron spectroscopy
XRD x-ray diffraction
YIG yttrium iron garnet
ZTA zirconia-toughened alumina

Table of Contents

A

α-Al$_2$O$_3$ Pure alumina. *Polycrystalline* Al$_2$O$_3$ is known as corundum and *single crystals* as sapphire. Its *crystal structure* can be described as consisting of two *sublattices*: an *FCC* sublattice of O^{2-} ions and a sublattice of Al^{3+} ions occupying two thirds of the *octahedral sites* in the first one.

α-Fe *Allotropic form* of iron having *BCC crystal structure* and existing at temperatures below 910°C at atmospheric pressure.

α isomorphous Ti system Ti–X *alloy system* in which the *alloying element X* is the *α-stabilizer*, i.e., it raises the temperature of the *β ↔ α polymorphic transformation*.

α-phase [in Ti alloys] A solid solution of alloying elements in α-Ti.

α'-martensite See *titanium martensite*.

α"-martensite See *titanium martensite*.

α-stabilizer In physical metallurgy of Ti alloys, an *alloying element* increasing the *thermodynamic stability* of α-phase and thereby raising the β/(α + β) *transus* in the corresponding *phase diagram*. In physical metallurgy of *steels*, it is referred to as ferrite-stabilizer.

α-Ti *Allotropic form* of titanium having a *hexagonal crystal structure* and existing at temperatures below 882°C at atmospheric pressure. The *axial ratio* of its *lattice c/a* = 1.58, i.e., a little smaller than in an ideal *HCP structure*.

α Ti alloy Titanium alloy in which α-*phase* is the only *phase constituent* after *air-cooling* from the β-field in the *phase diagram* concerned. Alloys with a small fraction of β-*phase* (~5 vol%) are usually related to the same group and are called near-α alloys. All the α alloys contain α-*stabilizers*.

(α + β) brass Brass with two *phase constituents*: a copper-based *substitutional solid solution* (α-phase) and an *electron compound* (β-phase).

(α + β) Ti alloy Alloy whose *phase constituents* are α- and β-phases after *air-cooling* from the (α + β)-field in the *phase diagram* concerned. Slow cooling of these alloys from the β-field results in a *microstructure* comprising *grain-boundary allotriomorphs* of the α-phase (known as "primary" α) and packets of similarly oriented α-platelets with the β-phase layers between the platelets.

A$_1$/Ae$_1$ temperature In the Fe–Fe$_3$C diagram, the temperature of an *eutectoid reaction* corresponding to the PSK line in the diagram. Since the reaction, on cooling, starts at a certain *undercooling* (see *nucleation*), the temperature of its commencement, Ar_1, is lower than A_1. The start temperature of the same reaction on heating, Ac_1, is greater than A_1 due to *superheating*. The difference between Ac_1 and Ar_1 is named thermal, or transformation, hysteresis.

A₂/Ae₂ temperature Temperature of *magnetic transformation* in *ferrite* (~770°C). See *Curie temperature*.

A₃/Ae₃ temperature In Fe–Fe₃C *phase diagram*, a temperature of the *polymorphic transformation* $\gamma \leftrightarrow \alpha$ corresponding to the GS line in the diagram. *Critical points* on cooling and heating are known as Ar_3 ($Ar_3 < A_3$) and Ac_3 ($Ac_3 > A_3$), respectively. For details, see *A_1 temperature*.

A₄/Ae₄ temperature In Fe–Fe₃C *phase diagram*, a temperature of the polymorphic transformation δ-ferrite $\leftrightarrow$ austenite.

A$_{cm}$/Ae$_{cm}$ temperature In Fe–Fe₃C *phase diagram*, a temperature corresponding to the equilibrium *austenite $\leftrightarrow$ cementite*; it is shown by the ES line in the diagram.

aberration Defect observed in *optical* and *electron microscopes*. It reveals itself in a colored (in optical microscopy) or slightly eroded or distorted image. The main types of aberration are: *chromatic, spherical*, distortion, *astigmatism*, and coma.

abnormal grain growth (AG) *Grain growth* wherein the *mean grain size* changes slowly at first, then, after a certain *incubation period*, increases abruptly, almost linearly, with time. Only a minority of the grains (~10^{-5}) grow in the course of abnormal grain growth. These grains can reach the size of several mm, whereas the *matrix* grains retain its initial size of several µm until it is consumed. The reason why the small grains cannot grow or grow slowly is retardation of their boundary migration by various *drag forces* as, e.g., by *grain-boundary solute segregation* (also known as *impurity drag*), by small precipitates (see *particle drag*), or by *thermal grooves* in thin films and strips (see *groove drag*). The matrix can also be stabilized by low *mobility* of the majority of *grain boundaries*, characteristic of materials with a strong *single-component texture*. The grains growing in the course of AG differ from the matrix grains by an increased *capillary driving force* owing to their increased initial size (see *normal grain growth*). Sometimes, their growth can be supported by a *surface-energy driving force* or by a driving force owing to decreased *dislocation density* (see *strain-induced grain boundary migration*). Time dependence of the volume fraction of abnormally large grains is similar to that of *primary recrystallized* grains; owing to this, AG is often referred to as secondary recrystallization. In some cases, AG is quite helpful, as, e.g., in electrical steels, where it leads to the *Goss texture* formation and to a significant improvement in magnetic properties. In other cases, it is detrimental, as, e.g., in *crystalline ceramics* (see *solid-state sintering*). AG is also termed discontinuous or exaggerated grain growth.

abnormal pearlite In *hypereutectoid steels*, a *microstructure* formed by *pearlite colonies* separated by extended *ferrite* fields from the network of *proeutectoid cementite*.

absorption Phenomenon of taking up atoms or energy from the environment into a body. A reduction in the intensity of certain radiation passing through a substance is described by an *absorption coefficient*.

absorption coefficient Quantity describing a reduction of the *integrated intensity* of some radiation passed through a homogeneous substance. See *linear absorption coefficient* and *mass absorption coefficient*.

absorption contrast Image contrast associated with different x-ray (electron) *absorption* in the sample areas having different thicknesses or densities. It is also known as amplitude contrast.

absorption edge See *x-ray absorption spectrum*.

absorption factor Quantity characterizing an angular dependence of the intensity of diffracted x-ray radiation, the dependence being a result of the x-ray *absorption*. The absorption factor can increase with the *Bragg angle*, as e.g., in the *Debye-Scherrer method*, or remain independent of it, as e.g., in the *diffractometric method*. Absorption factor is taken into account in *x-ray structure analysis*.

absorption spectrum *Wavelength spectrum* of an absorbed radiation.

acceptor *Dopant* in semiconductors increasing the concentration of charge carriers. The energy level of the acceptor valence electrons lies within the *band gap* close to its bottom. Owing to this, valence electrons from the filled *valence band* can be activated to the acceptor level, which, in turn, produces empty levels (known as *holes*) in the valence band, and thus promotes the electron conductivity. For instance, in elemental semiconductors (Si, Ge), acceptors can be *substitutional solutes* with a smaller valence than that of *host* atoms.

accommodation strain See *coherency strain*.

achromatic lens/objective In *optical microscopes*, a lens corrected for *chromatic aberrations* in two colors (usually red and green), as well as for *spherical aberrations*.

acicular Needle-shaped. The name has its origin in the fact that plate-like crystallites, as e.g., *Widmannstätten ferrite* or *steel martensite*, look like needles on plane sections studied by *optical microscopy*, *PEEM*, and *SEM*.

acicular ferrite Ferrite *crystallite* growing, apparently, as in the course of *bainitic transformation*. It has a lath-like shape and an increased *dislocation density*. The lathes form packets in which they are parallel to each other, and the boundaries between them inside a packet are *low-angle*. Several packets can occur within an *austenite grain*. Acicular ferrite is also termed Widmannstätten ferrite.

acicular martensite *Crystallite* of martensite in *steels* with a low M_s temperature of a lens- or needle-like shape in the cross-section. Martensite plates have a clearly visible longitudinal center line called midrib (i.e., middle ribbon). An increased density of *transformation twins* and *dislocations* is observed close to the midrib. The adjacent martensite plates of acicular martensite are non-parallel. The *habit planes* of acicular martensite are $\{259\}_A$ or $\{3\ 10\ 15\}_A$, and its *lattice* is oriented with respect to the *austenite* lattice according to the *Nishiyama* and *Greninger–Troiano orientation relation-*

ships, respectively. Acicular martensite is also called lenticular or plate martensite.

Ac temperature In Fe–Fe$_3$C *alloys,* a *critical point* observed on heating and denoted by Ac$_1$, Ac$_3$, or Ac$_{cm}$, for A_1, A_3, or A_{cm}, respectively. See *superheating.*

activation analysis Technique for chemical analysis wherein a sample is preliminary irradiated, and a *secondary radiation* of some component is used for determining its amount.

activation energy Additional *free energy* necessary for the commencement of some *thermally activated* reactions (e.g., *diffusion, recrystallization, phase transformations,* etc.). If activation energy is denoted by H, the *Gibbs free energy* is implied (in this case, activation energy can be referred to as activation enthalpy). If not, either the *Gibbs* or *Helmholtz free energy* may be meant. Units of activation energy are J/mol or eV/at.

activation enthalpy See *activation energy.*

active slip system Slip system over which the *dislocation glide* motion takes place.

adatom Atom from the environment adsorbed at the surface of an *adsorbent.*

adiabatic approximation The assumption that all processes in a *system* proceed without heat exchange with the environment.

adsorbate See *adsorption.*

adsorbent See *adsorption.*

adsorption *Spontaneous* attachment of atoms (or molecules) of some substance from the environment to the surface of some body, the substance being called *adsorbate* and the body *adsorbent.* Adsorption is accompanied by a decrease of surface energy. Adsorption results in the formation of an adsorption layer in which the adsorbate *concentration* is greater than in the environment. A layer of this kind can also form at some *lattice defects,* such as *grain boundaries* and *interfaces,* the environment and adsorbate being the bulk of the grains and *solute* atoms, respectively. In this case, adsorption is referred to as *equilibrium segregation.* See also *physical adsorption* and *chemisorption.*

after-effect Any alteration evolving after the completion of an external action.

age hardening An increase in hardness and strength caused by *precipitation treatment* resulting in precipitation of dispersed phase(s) from a *supersaturated solid solution.* It is frequently referred to as *precipitation strengthening.*

aging *Decomposition* of a *supersaturated solid solution.* The size and number of *precipitates* depends on the aging temperature and time and on the *supersaturation,* as well as on the solution *substructure* (see *heterogeneous nucleation*). Their arrangement is affected by the *microstructure* of the supersaturated solution and the previously mentioned aging conditions. For instance, if precipitates nucleate and grow inside the parent grains, *Widmannstätten structure* can appear. If they nucleate and grow predominately at the *subboundaries* and *grain boundaries* of the parent phase, the precipitates can form a network corresponding to the boundary network of the parent phase. In addition, narrow *precipitation-free zones* near the grain boundaries can occur.

aging [in Ti alloys] *Phase changes* accompanying the *decomposition* of *retained* β-*phase* or *metastable* β-*phase* (β_m) that occurred on *tempering*. These changes are commonly referred to as aging, although both β and β_m are *unsaturated* with respect to the *equilibrium* β-phase at the aging temperature. In the course of aging, phases with a decreased *solute concentration* precipitate from the metastable β-phases. As a result, the latter become solute-rich and their *composition* tends to the equilibrium β-phase. Possible sequences of phase changes during aging in (α + β) *alloys* can be described as follows: $\beta_m \rightarrow \beta_m + \omega \rightarrow \beta + \alpha$ or $\beta_m \rightarrow \beta_1 + \beta_2 \rightarrow \beta + \alpha$. Here, β_1 and β_2 are *metastable BCC* phases differing in composition from β_m and from one another; they supposedly occur by *spinodal decomposition* of β_m. The *microstructure* after aging consists of two *microconstituents*: β-*matrix* and relatively uniformly distributed, dispersed α-*phase* particles. See ω-*phase* and *aging treatment*.

aging treatment *Heat treatment* aimed at *age hardening*; it comprises *solution* and *precipitation treatments*.

aging treatment of Ti alloys *Heat treatment* that comprises heating of quenched alloys with *metastable* α′-, α″-, and ω-*phases* and *retained* β- or *metastable* β-*phase*. As for α′-*martensite*, α″-*martensite*, and ω-phase, the treatment should be named *tempering*, whereas the term "aging treatment" should relate solely to the previously mentioned β-phases. See *aging in Ti alloys* and *tempering of titanium martensite*.

air-cooling Cooling in still air.

aliovalent solute/impurity *Solute* in *ionic crystals* whose valence differs from that of a *host* ion. Aliovalent solutes disturb the electrical neutrality and must be associated with other defects (either *lattice defects* or *electrons*) compensating their charge.

allotropic change Transformation of one *allotropic form* into another, the transformation evolving as a *first-order transition*. See also *polymorphic transformation*.

allotropic form/modification In a single-*component* solid, one of several *stable phases* differing from the others by *crystal structure*, and transforming one into another *spontaneously* at the corresponding temperature and pressure. There can be more than two allotropic forms. They are usually denoted by Greek letters in alphabetic order, starting with alpha for the lowest temperature form. See *allotropy*.

allotropy In a single-*component* solid, the existence of *stable phases* with different *crystal structures* in different temperature or pressure ranges. Allotropic transformation relates to *first-order transitions*. See also *polymorphism*.

alloy Metallic material consisting of a *base* metal and one or more *alloying elements* partially or completely dissolving in the base metal. Alloys are frequently denoted by symbols of their *components*, the symbol of the base metal being usually underlined, as, e.g., <u>Cu</u>–Zn alloy for *brasses*.

alloy carbide *Intermediate phase* in *alloy steels* consisting of carbon and *alloying element(s)*. It is also termed special carbide.

alloying composition Auxiliary alloy used in the *alloy* production instead of pure *alloying elements*. It is also known as master alloy.

alloying element *Component* added deliberately with the aim of improving the properties of an *alloy*. Alloying elements can affect the existence range of *equilibrium phases* present in an unalloyed material, or lead to the occurrence of new phases, or both. In addition, alloying elements strongly affect the *kinetics* of *phase transformations* and thus the *microstructure* formation in alloyed materials. See also *dopant*.

alloy steel Steel comprising one or several *alloying elements*, along with carbon.

alloy system See *system*.

alpha brass Brass with only one *phase constituent*, that is, a Cu-based *solid solution*.

ambipolar diffusion Coupled migration of oppositely charged ions and *lattice defects* under the influence of an electric field, either external or internal. In the latter case, the oppositely charged species migrate together because their separate migration disturbs the electrical neutrality. Ambipolar diffusion may be observed in *sintering* and *diffusional creep* of *ionic crystals*. Compare with *electromigration* in metals.

amorphous solid Phase characterized only by a *short-range order* and by a missing *long-range order* in *atomic structure*. Amorphous phase can be obtained by *quenching* the melt below a *glass transition temperature* (see *glassy phase*), by *ion bombardment*, by heavy *plastic deformation* (e.g., by *mechanical alloying*), by rapid film deposition, etc.

amplitude contrast See *absorption contrast*.

analytical electron microscope (AEM) *TEM* used for chemical analysis of small areas (~10 nm in diameter), e.g., by means of *EELA*.

Andrade creep *Transient creep* described by the empirical time dependence of the creep *strain,* ε, in tension tests:

$$\varepsilon \propto at^{1/3}$$

(*a* is a constant and *t* is time). Andrade creep is observed at higher temperatures than *logarithmic creep*.

anelasticity Deviation from the behavior according to *Hooke's law* that reveals itself in two constituents of *elastic strain*: an instantaneous one, occurring simultaneously with the application of an external force and corresponding to the Hooke law; and a time-dependent constituent, $\varepsilon(t)$, changing with time after the force application, *t*, at a constant temperature as follows:

$$\varepsilon(t) = (\sigma/E)[1-\exp(-t/\tau_R)]$$

where *E* is *Young's modulus*, σ is the *tensile stress*, and τ_R is the relaxation time. The relaxation time is constant at a fixed σ and a constant temperature and is dependent on the nature of anelasticity. As seen in the equation, at various t, there may be different values of the *elasticity modulus*, its extremities being the Young's modulus E corresponding to the Hooke's law at $t \ll \tau_R$ and what is known as the relaxation modulus $E_R < E$ at $t \gg \tau_R$. See also *internal friction*.

anisotropic Having different *physical* and *mechanical properties* in various directions. Anisotropy of *single crystals* is a result of *crystalline anisotropy*, whereas that of a *polycrystal* is dependent on *crystallographic texture* (and so on the crystalline anisotropy) as well as on the microstructural anisotropy as, e.g., *banded structure* or *carbide stringers* in *steels* or an elongated *grain structure* in heat-resistant alloys (see *Nabarro–Herring* or *Coble creep*). Anisotropy can be observed not only in crystalline solids but also in some liquids (see *liquid crystals*).

annealing/anneal *Heat treatment* resulting in the occurrence of *equilibrium phases* (see, e.g., *graphitization anneal, solution annealing*), in removing of deformation or amorphization effects or in attaining a required *grain size* or *texture* (see, e.g., *recrystallization annealing*), or in relieving chemical inhomogeneity and *macroscopic residual stresses* (see *homogenizing, stress-relief annealing*). In metallic alloys, annealing is a preliminary treatment preparing the *microstructure* or *phase composition* to a final treatment (see, e.g., *austenitization* and *solution treatment*). Annealing after amorphization of *single-crystalline* semiconductors can restore single-crystalline structure.

annealing texture *Preferred orientation* evolved in the course of *primary recrystallization* or *grain growth*. *Recrystallization texture* occurs because *recrystallization nuclei* are of nonrandom orientations and grow into the deformed matrix at different rates. It can be similar to *deformation texture* or quite different from it. Texture changes during grain growth are connected with different *driving forces* for growth of variously oriented grains and different *mobility* of their boundaries (see *compromise texture*). Grain growth commonly (but not always) results in weakening of the primary recrystallization texture. Annealing texture is usually characterized by an increased *scatter* and a decreased *intensity* in comparison to the initial deformation texture, except for a *cube texture* in some cold-rolled *FCC alloys* and the *Goss texture* in *ferritic steels*.

annealing twin *Twin* occurring during *primary recrystallization* or *grain growth*. Annealing twins are usually observed in materials with low *stacking-fault energy*, especially on *annealing* after heavy *plastic deformation*. An annealing twin, depending on its position inside a *grain*, can have one or two *coherent twin boundaries* joining up with *grain boundaries* or *incoherent twin boundaries*. The twin with two coherent boundaries looks like a straight band.

anomalous x-ray transmission Abnormally low x-ray *absorption* observed in thick *perfect crystals* adjusted at the exact *Bragg angle*. It is also known as the Borrmann effect.

antiferromagnetic Material characterized (below *Néel point*) by a negative energy of *exchange interaction* and equal but oppositely directed magnetic moments of different magnetic *sublattices*. The intrinsic magnetization in antiferromagnetics is lacking due to equality of the magnetic moments of the sublattices.

antiferromagnetic Curie point See *Néel point*.

antiphase boundary Boundary of *antiphase domains* within a *grain* of an *ordered solid solution*. Antiphase boundary is characterized by an increased energy because the arrangement of atoms of different *components* at the boundary is distorted in comparison to their arrangement inside domains (see Figure A.1).

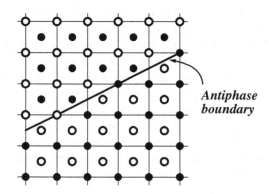

FIGURE A.1 Antiphase domains and an antiphase boundary inside a grain of an ordered solid solution. Open and solid circles represent atoms of different components.

antiphase domain *Grain* part having a *crystal structure* of an *ordered solid solution*. Identical *sublattices* in the adjacent antiphase domains inside one grain are shifted relative to each other (see Figure A.1), the shift being unequal to the *translation vector* of the corresponding *superlattice*. If the superlattice is of a noncubic *system*, identical sublattices of the adjacent antiphase domains inside a grain can have different spatial orientations.

antisite defect *Lattice defect* in *ionic crystals* produced by an ion of some sign occupying a site in the *sublattice* formed by ions of the opposite sign. Antisite defect is analogous to an *antistructural atom* in *metallic crystals*. See *structural disorder*.

antistructural atom See *structural disorder*.

aperture diaphragm In *optical microscopes*, a diaphragm that restricts the incident beam and affects the illumination intensity, image contrast, *resolving power*, and *depth of focus*.

apochromatic lens/objective In *optical microscopes*, a lens corrected for *chromatic aberration* in three color regions (violet, green, and red) and for *spherical aberration* in two color regions (violet and green). Apochromatic objective has a better color correction than *achromatic objective*.

arrest point See *critical point* and *thermal analysis*.

Arrhenius equation Description of the temperature dependence of some kinetic parameter, A, of any *thermally activated* process:

$$A = A_0 \exp (-Q/cT)$$

Here, A_0 is a pre-exponential factor, Q is the *activation energy*, T is the absolute temperature, and c is either the *gas constant* (if the activation of one molecule is considered) or the *Boltzmann constant* (if the activation of one atom, or molecule, is concerned).

Ar temperature In Fe–Fe$_3$C *alloys*, a *critical point* observed on cooling and denoted by Ar_1, Ar_3 or Ar_{cm}, for A_1, A_3, or A_{cm}, respectively. See *undercooling*.

artifact Feature caused by preparation or manipulation of a sample or, sometimes, by investigation conditions.

artificial aging *Aging treatment* at temperatures higher than ambient.

asterism Radial elongation of reflection spots in Laue diffraction patterns owing to *residual stresses* or *substructures* in a *single crystal*.

astigmatism Optical *aberration* revealing itself in a distortion of the cylindrical symmetry of an image.

asymmetric boundary *Tilt grain boundary* whose plane divides the angle between identical planes in the lattices of the adjacent *grains* into two unequal parts.

athermal transformation *Phase transition* developing without any *thermal activation* (thus, the transformation is *diffusionless*). The volume fraction of the transformation products depends mostly on temperature (or, more precisely, on *supercooling*). At a fixed temperature in the *transformation range*, after some period of a rapid increase, the volume fraction changes little, if at all. See *shear-type transformation* and *martensitic transformation*.

atomic force microscope (AFM) Device for studying the surface *atomic structure* of solids. AFM is similar in design to *STM*, but measures the force between the sharp microscope tip and surface atoms.

atomic mass Atomic mass, in units, equal to 1/12 mass of ^{12}C atom.

atomic packing factor Volume fraction of a *unit cell* occupied by atoms presented by rigid spheres of equal radii. The largest atomic packing factor is 0.74 in *FCC* and *HCP lattices*; it is a little smaller (0.68) in *BCC* lattice, and very low (0.34) in the *diamond* lattice. Atomic packing factor is also known as packing factor.

at% Atomic percentage; it is used in cases in which the *components* are chemical elements. A *weight percentage* of a component A, W_A, in a *binary system* A–B can be calculated from its atomic percentage, A_A, by the formula:

$$W_A = 100/[1 + (100 - A_A)M_B/(A_A M_A)]$$

where M_A and M_B are the *atomic masses* of A and B, respectively. In cases in which the components are compounds, *mol%* is used instead of at%.

atomic radius Conventional value not connected with an atomic size, but relating to a *crystal lattice,* i.e., the *interatomic spacing* is assumed equal to the sum of atomic radii. This is the reason why atomic radius depends on the bond type (i.e., *metallic, ionic,* or *covalent*), as well as on the *coordination number* in the *crystal lattice* considered. See *metallic, ionic,* and *covalent radii.*

atomic scattering factor Coefficient characterizing the intensity of the *elastically scattered* radiation. It increases with the atomic number and decreases with $(\sin \theta)/\lambda$, where θ is the *glancing angle* and λ is the wavelength. The atomic scattering factor for electrons is $\sim 10^5$ times greater than for x-rays, which enables the application of *electron diffraction* for studying relatively thin objects, commonly of thickness smaller than 0.1 μm. Atomic scattering factor is taken into account in *x-ray structure analysis.*

atomic size See *atomic radius.*

atomic structure In materials science, a description of an atomic arrangement in *phases,* e.g., *amorphous,* or in *lattice defects.*

atomic volume Volume of *unit cell* per atom.

atomizing Procedure for obtaining small solid droplets from melt, the droplets being *ultra-fine grained* because the cooling rate during their *solidification* is $\sim 10^3$ K/s. They are used for producing massive articles by consolidating and *sintering.*

atom probe field ion microscopy (APFIM) Technique for mass-spectrometric identification of single atoms removed from the sample tip in *FIM* by means of pulse field evaporation. Besides the studies of the surface *atomic structure,* APFIM is used for analyzing the *nucleation* and *growth* of *precipitates,* ordering phenomena (see *order–disorder transformation* and *short-range ordering*), and *segregation* at *crystal defects.*

Auger electron *Secondary electron* emitted by an atom whose electron vacancy at an inner shell has been created by a high-energy primary electron. An electron from a higher energy shell subsequently fills the electron vacancy, whereas another electron, referred to as the Auger electron, is emitted from the other shell. The *energy spectrum* of Auger electrons is a characteristic of the atom and can be used for chemical analysis (see *Auger-electron spectroscopy*).

Auger-electron spectroscopy (AES) Technique for chemical analysis utilizing the *energy spectrum* of *Auger-electrons.* Since Auger-electrons are of low-energy, AES can analyze very thin surface layers only (~1 nm in depth), with the lateral resolution 20 to 50 nm. AES can also yield a depth profile of chemical *composition* using *ion etching* for the layer-by-layer removal of the material studied.

ausforming *Thermo-mechanical treatment* comprising two main stages: *warm deformation* of a *steel* article at temperatures of *bainitic range* for the time period smaller than the *incubation period* of *bainitic transformation;* and *quenching* of the article, which results in the *martensite* or *bainite* formation from the deformed *austenite.* An increased *dislocation density*

in the austenite (after the first stage) is inherited by the martensite or
bainitic *ferrite* (after the second stage), which increases the article's *hardness*. Ausforming is also referred to as low-temperature thermo-mechanical treatment.

austempering *Heat treatment* comprising *austenitization* of a *steel* article, cooling it to a *bainitic range* at a rate higher than the *critical cooling rate* and
holding at a fixed temperature until the completion of *bainitic transformation*.

austenite *Solid solution* of *alloying elements* and/or carbon in γ-*Fe*. It is named
after British metallurgist W. C. Roberts-Austen.

austenite finish temperature (A_f) Temperature at which the transformation of
martensite into *austenite* completes upon heating. The same designation
is also applied to nonferrous alloys in which martensite transforms into
some parent phase.

austenite stabilization Decrease, in comparison to a continuous cooling, in the
amount of *martensite* occurring from *austenite* when cooling is interrupted
at a temperature between M_s and M_f. This can be explained by the relaxation of stresses induced in the austenite by martensite crystals occurring
before the interruption. The relaxation, in turn, leads to the *dislocation*
rearrangement and their interaction with martensite/austenite *interfaces*,
which makes the interfaces immobile.

austenite-stabilizer *Alloying element* expanding the γ-phase field in the corresponding *phase diagram*, which manifests itself in a decrease of the A_3
temperature and an increase of the A_4 temperature in *binary* alloys Fe–M
as well as in a decrease of A_1 temperature in *ternary* alloys Fe–C–M (M
is an *alloying element*). The solubility of austenite-stabilizers in *ferrite* is
much lower than in austenite. Under the influence of austenite-stabilizers,
austenite can become thermodynamically stable down to room temperature. See, e.g., *austenitic steels*.

austenite start temperature (A_s) Temperature at which the transformation of
martensite into austenite starts upon heating. The same designation is also
applied to nonferrous alloys in which martensite transforms into some
parent phase.

austenitic-ferritic steel *Alloy steel* whose *structure* after *normalizing* consists
of *austenite* and *ferrite*.

austenitic-martensitic steel *Alloy steel* whose *structure* after *normalizing* consists of *austenite* and *martensite*.

austenitic range Temperature range wherein a purely austenitic *structure* can
be obtained in *steels* upon heating.

austenitic steel *Alloy steel* whose *structure* after *normalizing* consists predominately of *austenite*. This is a result of an increase in the thermodynamic
stability of austenite by *alloying elements*. If austenite is thermodynamically unstable, it can transform into *martensite* (see, e.g., *maraging steel*
and *transformation-induced plasticity*).

austenitization Holding stage of a *heat treatment* resulting in the formation of
a completely *austenitic structure*.

autoelectronic emission See *field emission*.

autoradiography Technique for studying chemical microinhomogeneity by registering the radiation of radioactive elements (tracers) contained in the specimen on a high-resolution screen (film), displaying the disposition of the tracers in the surface layer.

Avogadro number Amount of atoms, ions, or molecules in a mole of any substance; $N_A \cong 6.022 \cdot 10^{23}$ mol^{-1}.

Avrami equation Description of transformation *kinetics,* assuming that new phase *nuclei* occur at predetermined sites only. As a result of this assumption, the *nucleation rate* decreases with time. In this case, the kinetic equation is:

$$V/V_0 = 1 - \exp(-kt^n)$$

where V and V_0 are the transformed and the initial volume fractions, respectively, k is a kinetic constant, t is the transformation time, and $3 \leq n \leq 4$ (in three-dimensional cases) or $2 \leq n \leq 3$ (in two-dimensional cases). If the nucleation rate is constant, the Avrami equation is identical to the *Johnson–Mehl–Kolmogorov equation*. In cases in which all the nucleation sites are exhausted at an early stage:

$$V/V_0 = 1 - \exp[-(4\pi N_0/3)G^3 t^3]$$

where N_0 is the initial number of the nucleation sites and G is the *linear growth rate*.

axial angle In a *unit cell*, an angle between a pair of its axes. See *lattice parameters* and *unit cell* (Figure U.2).

axial ratio In hexagonal crystal systems, the ratio of *lattice constants* c and a.

B

β-Al₂O₃ Impure alumina whose main impurity is Na_2O.

β **eutectoid Ti system** Name of a Ti–X *alloy system* in which the β-*stabilizer* X has a *limited solubility* in β-*Ti*, and a *eutectoid reaction* β ↔ α + γ takes place (γ is an *intermediate phase* or a *terminal solid solution*).

β-**Fe** Obsolete designation of the *paramagnetic* α-*Fe* existing at temperatures between 768 and 910°C at atmospheric pressure (i.e., between A_2 and A_3). Correspondingly, a *solid solution* in β-*Fe* was named β-ferrite.

β **isomorphous Ti system** Name of a Ti–X *alloy system* in which the *alloying element* X is the β-*stabilizer* and there is no *eutectoid reaction* in the corresponding *phase diagram*.

β_m **phase [in Ti alloys]** See *metastable* β-*phase*.

β-**phase [in Ti alloys]** Solid solution of alloying elements in β-Ti.

β-**stabilizer** *Alloying element* expanding the β-phase field in *phase diagrams* of Ti alloys and thereby lowering β/(α + β) *transus*.

β-**Ti** High-temperature *allotropic form* of titanium having *BCC crystal structure* and existing above 882°C up to the *melting point* at atmospheric pressure.

β **Ti alloy** Alloy with β-*stabilizers* wherein β-phase is the only *phase constituent* after *air-cooling* from temperatures above the β/(α + β) *transus*. Alloys with a small (~5 vol%) amount of α-*phase* are related to the same group and termed near-β alloys. If the β → α transition does not evolve on air-cooling, these alloys are named *metastable* β *alloys*.

background In *x-ray structure analysis* and *texture analysis*, an intensity of scattered x-ray radiation between *diffraction lines* caused mainly by: *x-ray fluorescent radiation* emitted by the specimen, diffraction of the *white radiation* on the *polycrystalline* specimen, *Compton scattering*, and *diffuse scattering*.

back-reflection Laue method Technique wherein an x-ray source and a flat film (screen) registering an x-ray diffraction pattern are placed on the same side of the sample.

backscattered electron Electron *elastically scattered* in the direction that is opposite to the direction of the primary beam. The yield of backscattered electrons increases with the atomic number of the substance studied. Backscattered electrons are used in *SEM* for gaining data on the topography, *microstructure,* and chemistry of the specimen surface, as well as for crystallographic studies (see *electron channeling*).

bainite *Microconstituent* in *steels* occurring on transformation of *undercooled austenite* in a *bainitic range*. Bainite consists of *ferrite* and *cementite* (or *ε-carbide*). It is named after American scientist E. C. Bain. See *bainitic transformation, upper bainite,* and *lower bainite.*

bainite start temperature (B_s) In *alloy steels*, temperature of the start of *bainitic transformation* on cooling from an *austenitic range.*

bainitic range Temperature range wherein *bainite* can be obtained upon cooling from an *austenitic range*. The upper limit of bainitic range is the B_s temperature in *alloy steels* and the lower limit of *pearlitic range* in *plain carbon steels*. The lower limit of the bainitic range is the M_s temperature (see Figure B.1).

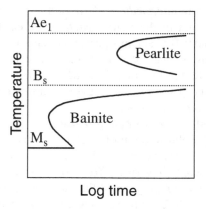

FIGURE B.1 TTT diagram for eutectoid alloy steel (scheme). Temperature range Ae_1–B_s is referred to as pearlitic, B_s–M_s as bainitic, and M_s–M_f as martensitic (temperature M_f is not shown).

bainitic transformation In *steels, phase transformation* of *undercooled austenite* at temperatures of *bainitic range*. In this range, the atoms of both iron and *substitutional alloying elements* cannot migrate by *diffusion*, whereas the carbon atoms can. Bainitic transformation (BT) evolves as follows. Carbon diffusion inside *austenite* leads to its chemical inhomogeneity, i.e., in some areas, the carbon content becomes reduced and in the others, increased. Since the M_s increases with a reduction of the carbon *concentration* in austenite, *martensitic transformation* evolves in the low-carbon areas. An occurring *metastable* low-carbon *martensite* decomposes into *ferrite* because of the elimination of its carbon content through the *carbide* precipitation. If this proceeds into the upper part of bainitic range, diffusion paths of carbon atoms can be long enough, and the carbides occur only at the boundaries of the ferrite crystallites (see *upper bainite*). If the temperature is low, the diffusion paths are short, and the carbides form inside the ferrite grains (see *lower bainite*). In the high-carbon areas, the austenite transforms into a ferrite-cementite mixture in the upper part of bainitic range. In the lower part of the range, carbide precipitation in these areas leads to a further local reduction of the carbon content in the austenite and to the transformation chain described. Sometimes, a certain part of the austenite remains untransformed (see *retained austenite*). Thus, BT includes a *diffusion-controlled* carbon partitioning inside austenite, a *nondiffusional phase transition* of austenite into martensite, and a diffusion-controlled precipitation of carbides from the martensite and austenite. The BT *kinetics* are governed by the slowest process, i.e., by the carbon diffusion, and are the same as in the other *diffusional transformations*. At the same time, BT is similar to martensitic transformation in the sense that it ceases continuously at a constant temperature below B_s, and a certain amount of austenite remains untransformed. *Alloying elements* affect BT by reducing the carbon *diffusivity* and changing the *elastic modulus* of austenite, which retards the transformation and lowers the B_s temperature.

bainitic steel *Alloy steel* whose *microstructure* after *normalizing* consists predominately of *bainite*.

bamboo structure *Microstructure* of thin wires formed by a row of *grains* whose diameter is equal to the wire diameter.

banded structure *Microstructure* of an article fabricated from *hypoeutectoid carbon steel,* wherein *pearlite* and *proeutectoid ferrite* form alternating bands parallel to the direction of the preceding *hot deformation*. Banded structure has its origin in the *coring* in a steel *ingot*. For instance, in silicon steels, proeutectoid ferrite occurs in the areas of the hot-deformed and dynamically *recrystallized* austenite where there is an increased silicon *concentration*, i.e., on the periphery of the prior *dendrite* arms. Banded structure leads to high *anisotropy* of the *mechanical properties* of steel articles.

band gap See *band structure*.

band structure Spectrum of available energy states for electrons in crystals. The spectrum is composed of almost-continuous bands of permitted energy states separated by the gaps of forbidden energy (these are called band gaps). The bond type and crystal structure determine the spectrum. The electrons of the upper-atom shell fill a valence band. The band of higher permitted energies, next to the valence band, is known as a conduction band; it can be completely or partially empty. Electron conductivity is only possible if valence electrons can be activated to the energy level corresponding to the conduction band. In metals, the valence and conduction bands lie close to each other or superimpose. This explains the high conductivity of metals in which there are always available energy states in the conduction band. In intrinsic semiconductors and insulators, the valence band is filled completely and the conduction band is empty, but the latter is separated from the former by a band gap. Thus, the electron conductivity in these materials is only possible if valence electrons of the highest energy can acquire the activation energy necessary to overcome the gap. In intrinsic semiconductors, this takes place under the influence of thermal, electrical, magnetic, or light excitation, because the band gap in these materials is relatively small. In intrinsic insulators, the band gap is large, so there is no electron conductivity in these materials. Certain impurities, known as donors and acceptors, introduce permitted energy levels into the band gap close to its borders, which reduces the activation energy necessary for electrons to reach the conduction band and significantly increases the number of charge carriers. Under the influence of some impurities, both covalent and ionic crystals can become semiconductors.

basal plane In crystallography, {0001} plane in *hexagonal structure*. See *Miller–Bravais indices*.

basal slip Slip over a basal plane along $\langle 11\bar{2}0 \rangle$ direction; it is commonly observed in *HCP* alloys with an axial ratio $c/a \geq 1.633$.

base In materials science, a *component* used as a basis for *alloying*.

base-centered lattice *Orthorhombic* or *monoclinic Bravais lattices* in which, along with the *lattice points* at the vertices of the corresponding *unit cell*, there are additional points at the centers of two opposite faces. It is also referred to as based lattice.

based lattice See *base-centered lattice*.

Bauschinger effect In the specimen strained initially in one direction and then in the reverse direction, a decrease of the *yield stress* observed on the second loading. *Microscopic residual stresses* induced upon the first loading cause this effect due to inhomogeneity of plastic flow.

bend contour See *extinction contour*.

bicrystal Solid body consisting of only two *crystallites* of the same or different *phases*. The latter case is usual in semiconductor *heterojunctions* wherein bicrystal is formed by a *heteroepitaxial single-crystalline* film on a single-crystalline substrate or by two single-crystalline films. If bicrystal consists of crystallites of the same phase, they are *disoriented*.

bimetallic Consisting of two brazed or welded metallic strips of different composition and properties. For instance, thermobimetals are produced from strips with different coefficients of thermal expansion.

bimodal Description of a curve with two distinct maxima.

binodal Dome-shaped surface or a curve in a *ternary phase diagram* and a *binary* diagram, respectively, bordering a *miscibility gap* (see Figures B.2 and B.3).

binary Consisting of two *components*.

black-heart malleable [cast] iron *Malleable iron* with a *pearlitic matrix*.

Bloch wall *Domain wall* characteristic of massive *ferromagnetics* or *ferrimagnetics*. Inside the wall, the magnetization vector rotates around an axis perpendicular to the wall plane, going from one domain to the other. In thin films, such a wall structure is thermodynamically unfavorable (see *Néel wall*). The thickness of a 180° Bloch wall is proportional to $(A/K)^{1/2}$, where A is the energy of *exchange interaction* and K is the constant of *magnetic crystalline anisotropy*; e.g., Bloch-wall thickness equals ~50 nm in α-*Fe* and ~3 nm in an *intermediate phase* $Fe_{14}Nd_2B$.

blocky martensite See *lath martensite*.

body-centered cubic (BCC) structure *Crystal structure* whose *coordination number* equals 8; *atomic packing factor* is 0.68; the *close-packed planes* and the *close-packed directions* are {110} and $\langle 111 \rangle$, respectively (see Figure B.4); the radius of a *tetrahedral void* in the structure is 0.290R; and that of an *octahedral void* is 0.153R, where R is the *atomic radius*.

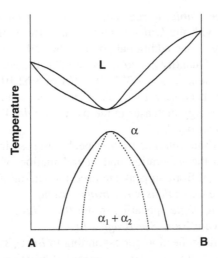

FIGURE B.2 Binary diagram with a miscibility gap in the solid state: α_1 and α_2 denote α solid solutions of different compositions. Solid line shows a binodal.

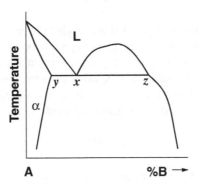

FIGURE B.3 Part of a binary phase diagram with a monotectic reaction. The dome-shaped curve is a binodal bordering $(L_1 + L_2)$ field in the case of monotectic reaction or $(\beta_1 + \beta_2)$ field.

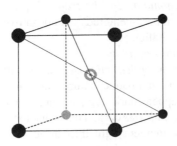

FIGURE B.4 Unit cell of BCC crystal structure. Open circle shows an atom inside the cell body.

body-centered lattice *Cubic, tetragonal,* or *orthorhombic Bravais lattices* in which, along with the *lattice points* at the vertices of the corresponding *unit cell*, there is one additional point at the cell's center.

Boltzmann constant Quantity $k = R/N_A$, where R is the *gas constant* and N_A is the *Avogadro number*: $k = 1.381 \cdot 10^{-23}$ J/K or $8.62 \cdot 10^{-5}$ eV/K.

bond energy Energy necessary to break interatomic bonds and separate the atoms. Bond energy increases in the order: *van der Waals* $\rightarrow$ *metallic* $\rightarrow$ *ionic* or *covalent bond.*

Bordoni peak/relaxation *Internal friction* peak observed in *cold-worked FCC* metals due to the generation and lateral motion of *double kinks*. The measurements of Bordoni peak are used for determining the kink energy.

Borrmann effect See *anomalous x-ray transmission.*

Bragg angle *Glancing angle* appearing in the *Bragg law.*

Bragg [diffraction] condition See *Bragg's law.*

Bragg reflection X-ray reflection corresponding to *Bragg's law.*

Bragg's law Condition for *x-ray* (or *electron*) *diffraction* on parallel *lattice planes* spaced at a distance d_{hkl}, where h, k, and l are the *Miller indices* of the planes:

$$n\lambda = 2\,d_{hkl}\,\sin\theta$$

where λ is the wavelength, θ is the angle between the primary beam and the corresponding planes (the glancing angle or Bragg angle), and n is an order of reflection, i.e., an integral value consistent with the condition $n\lambda/2d_{hkl} < 1$. It is assumed that the primary beam is strictly *monochromatic* and parallel, and that the crystal studied has a perfect lattice. The angle θ does not correspond to the angle of incidence, ϑ, considered in optics:

$$\theta = \pi/2 - \vartheta$$

This explains why θ is termed glancing angle.

brass Cu *alloy* where zinc is the main *alloying element.*

Bravais lattice One of 14 possible crystal lattices: *cubic* (primitive, body-centered, and face-centered); *tetragonal* (primitive and body-centered); *orthorhombic* (primitive, body-centered, base-centered, and face-centered); primitive *rhombohedral* or *trigonal*; primitive *hexagonal*; *monoclinic* (primitive and base-centered); and primitive *triclinic*. The names of *crystal systems* are italicized.

bremsstrahlung See *white radiation.*

bright-field illumination In *optical microscopy*, such illumination that flat horizontal features of an opaque sample appear bright, whereas all the inclined features appear dark; e.g., in single-phase materials, *grains* are bright and *grain boundaries* dark. This is due to the fact that the horizontal features reflect the incident light into an objective, whereas the inclined features do not.

bright-field image *TEM* image produced by a directly transmitted electron beam. Bright features in the image correspond to areas with an undistorted *lattice,* provided the image results from *diffraction contrast.*

bronze Cu-based *alloy* in which zinc is a minor *alloying element.* Bronzes are denoted by the name of the main alloying element as, e.g., aluminum bronze, silicon bronze, lead bronze, etc.

Bs/Def orientation One of the main *texture components,* {011}⟨211⟩, observed in cold-rolled *FCC* metallic materials of low *stacking-fault energy* as well as in the cold-rolled copper.

Bs/Rex orientation *Recrystallization texture component,* {236}⟨385⟩, observed in cold-rolled *FCC* metallic materials of low *stacking-fault energy.*

bulk diffusion Mass transport through the *grain* interiors in a *polycrystalline* material. It is also termed lattice diffusion or volume diffusion.

bulk modulus *Elastic modulus* at *hydrostatic pressure.* In non-*textured polycrystals*, it is *isotropic* and is usually denoted by K. Its magnitude relates to *Young's modulus, E,* as follows:

$$K = E/3(1 - 2\nu)$$

where ν is *Poisson's ratio.* See *Hooke's law.*

Burger orientation relationship Orientation relationship between an *HCP phase,* α, and a *BCC* phase, β: $\{0001\}_\alpha \| \{110\}_\beta$, $\langle 11\bar{2}0 \rangle_\alpha \| \langle 111 \rangle_\beta$.

Burgers circuit Closed circuit in a perfect *crystal lattice*; it helps determine the type of a *linear defect* in an imperfect crystal with the same *lattice.* If a circuit, identical to that in the perfect crystal, is drawn around a linear defect and turns to be opened, the defect is a *dislocation*; if it is closed, it is a *disclination.* The circuit is drawn counterclockwise around the defect. Thus, the defect *sense* should be chosen first.

Burgers vector Vector, **b**, invariant for a given *dislocation* line and characterizing the magnitude of *lattice distortions* associated with it (see *dislocation energy* and *dislocation stress field*). The sense of the Burgers vector is defined as follows: the end of the vector should be taken at the end of the *Burgers circuit,* and its head at the start point of the circuit. Burgers vector of a *perfect dislocation* is the *translation vector* in the *crystal structure* concerned. For example, the Burgers vector of perfect dislocations in *BCC structure* is $1/2 \langle 111 \rangle$, i.e., it lies along $\langle 111 \rangle$ direction and its length equals half the body diagonal, i.e., $b = 1/2\, a\sqrt{3}$ (*a* is the *lattice constant*). Burgers vector can be determined experimentally using *TEM.*

C

χ-carbide In high-carbon *steels*, a *transient phase* of the *composition* Fe_5C_2 with *monoclinic lattice*. It occurs upon *tempering* of as-quenched *martensite*.

CaF₂ structure *Crystal structure* wherein Ca^{2+} ions form an *FCC sublattice* and F^{1-} ions, occupying half the *tetrahedral sites* of the first one, form the second sublattice (see Figure C.1). CaF_2 is called *fluorite*.

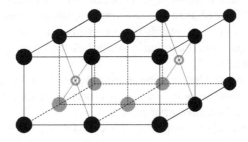

FIGURE C.1 Unit cell of CaF_2 crystal structure. Solid and open spheres show F^{1-} and Ca^{2+} ions, respectively.

calorimetry Technique for studying *phase transitions* by measuring thermal effects, i.e., taking off or releasing the heat in the course of the transitions. See, e.g., *differential scanning calorimetry*.

capillary driving force *Driving force* for migration of *grain* or *phase boundaries* under the influence of the boundary curvature (see *Gibbs–Thomson equation*); this driving force is directed to the center of the curvature. In a three-dimensional, single-phase *structure*, it is:

$$\Delta g = \gamma_{gb}(\rho_1 + \rho_2)$$

where γ_{gb} is the *grain-boundary energy*, and ρ_1 and ρ_2 are the principal radii of the boundary curvature. Capillary driving force promotes *normal* and *abnormal grain growth*, as well as *shrinkage* of porous compacts in the course of *sintering*.

carbide In *binary alloys*, an *intermediate phase* containing carbon. In alloys with more than two *components*, metallic components can dissolve in binary carbides, forming *complex carbides*.

carbide-former *Alloying element* able to form *special carbides.* In *steels,* it is always a transition metal situated in the groups of the Periodic Table of Elements to the left of iron.

carbide network In *hypereutectoid steels*, a continuous network of *proeutectoid* carbides precipitating upon slow cooling from *austenite* and arranging on its *grain boundaries.* Grain boundaries serve as preferred *nucleation* sites because the dissolved carbon segregates them (see *equilibrium segregation*) and the nucleation of a new phase proceeds easier at the boundaries rather than inside the grains (see *heterogeneous nucleation*).

carbide segregation An increased amount of eutectic and *proeutectoid* carbides in a certain area of an article resulting from *macrosegregation.*

carbide stringers Rows of coarse carbides along the direction of the preceding *hot deformation* of an article. They are observed in *ledeburitic steels,* in which small *eutectic colonies* containing a significant carbide fraction, after *solidification,* are arranged between *primary austenite grains.* In the course of the subsequent hot deformation, these grains elongate in the deformation direction, which is accompanied by the stretching of the colonies in the same direction. Carbides of the colonies are revealed as carbide stringers.

carbonitride *Carbide* in which carbon atoms are partially substituted by nitrogen atoms, or *nitride* in which nitrogen atoms are partially substituted by carbon atoms. See *interstitial phase.*

cast iron Iron-carbon *alloy* comprising 2.5–3.5 wt% C, ~1 wt% Mn, and 1–3 wt% Si, in which an *eutectic reaction* takes place during *solidification.*

casting Article obtained by pouring liquid metal into a hollow form, where it solidifies. Its *macrostructure* is similar to that of an *ingot,* however, usually without a *columnar zone.*

C-curve In a *TTT diagram,* a curve showing the development of *diffusional phase transformations* at different temperatures by depicting the time of the transformation start and, sometimes, the time of its finish (see Figure C.2).

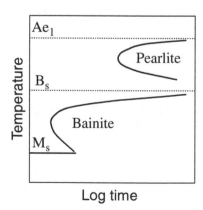

FIGURE C.2 TTT diagram for eutectoid alloy steel (scheme). Only C-curves of the transformation start are shown.

Long *incubation periods* at both low and high *supercooling* inside the corresponding *transformation range* are caused by a low *nucleation rate*, in the first case, and by a low *growth rate* in the second.

cell structure *Substructure* forming inside *cold deformed grains* at *true strains* $\gamma \geq 1$ (i.e., when the *stress-strain curve* becomes parabolic; see *strain hardening*). The cells are almost free of *dislocations* and bordered by *dislocation tangles* (known as cell walls) whose thickness is ~0.1 of the cell diameter. The mean cell diameter (usually ≤ 1 µm) varies as Gb/τ, where G is the *shear modulus*, b is the *Burgers vector*, and τ is the *shear stress*. Inside a grain, the mean cell *disorientation* increases from ~1° at $\gamma = 1$–2 to 5–10° at $\gamma = 5$–8°.

cellular microsegregation Inhomogeneous distribution of *solutes* in melt-grown *single crystals* with *cellular substructure*.

cellular precipitation See *discontinuous precipitation*.

cellular substructure *Substructure* in melt-grown *single crystals* characterized by pencil-like cells extending along the growth direction. These cells are delineated by areas with an increased *solute concentration* (known as *cellular microsegregation*), as well as by *low-angle boundaries*. Cellular substructure is observed in *dilute solid solutions*. Its occurrence is connected with *constitutional undercooling* of the melt next to the growth front. Cellular substructure is also referred to as lineage structure or striation structure.

cementite *Intermediate phase* of an approximate *composition* Fe_3C. Although cementite is *metastable*, it nucleates more rapidly than stable *graphite* because: its *nucleation* requires a smaller redistribution of carbon atoms; its *specific volume* is ~3 times smaller than that of graphite; and it can form a *partially coherent interface* with both *austenite* and *ferrite*, whereas the graphite/austenite and graphite/ferrite interfaces are always *incoherent* (see *critical nucleus* and *heterogeneous nucleation*).

characteristic x-rays X-rays of a unique wavelength (energy) characterizing the atoms in an x-ray source. Their wavelengths correspond to sharp-intensity maxima in a continuous *wavelength spectrum*. The shortest wavelengths (and thus, the highest energy) in the characteristic spectrum correspond to the lines of a K set, where the $K\alpha_1$ line is the most intensive one.

chemical diffusion Mass transport in *substitutional solid solutions* and *intermediate phases* in which different atoms diffuse either in the opposite directions or unidirectionally (see *ambipolar diffusion*); their diffusion rates are characterized by *intrinsic diffusivities*. Chemical diffusion is also referred to as interdiffusion.

chemical etching Treatment of polished *metallographic samples* by definite (usually *dilute*) chemical reagents for *microstructure* revealing. Etching results in the formation of a surface relief (i.e., surface roughening, grooves, or pits), making microstructural features visible. The surface roughening results from the dependence of the reaction rate on the orientation of the *grain* surface, whereas the grooves and *etch pits* occur owing to an increased reaction rate at the intersections of different *crystal defects*, such as *grain boundaries* and *interfaces,* as well as *dislocations*, with the sample surface.

chemical inhomogeneity *Composition* difference in various parts of an article (macroscopic inhomogeneity) or of a *grain* (microscopic inhomogeneity).

chemical potential. Partial derivative of *Gibbs free energy* with respect to the atomic (or molar) *concentration* of one *component* at constant concentrations of the other components, constant temperature, and constant volume.

chemisorption. *Adsorption* of atoms from the environment accompanied by the formation of *solid solutions* or *intermediate phases* on the surface of the solid *adsorbent*. In the absence of *diffusion*, the layer is of monoatomic (monomolecular) thickness. Chemisorption can be preceded by the dissociation of chemical compounds present in the environment.

cholesteric crystal. *Liquid crystal* wherein the rod-shaped molecules densely fill the space and form chains arranged end-to-end, as in *nematic crystals*. However, each chain in cholesteric crystals is helical, whereas in nematic crystals, it is almost straight-lined. Cholesteric crystals are also called N* crystals.

chromatic aberration. Aberration in *optical microscopes* appearing due to the difference in the focal lengths for light of different wavelengths. It reveals itself in colored image details.

cleavage plane. *Lattice plane* on which cleavage occurs as, e.g., {100} in *BCC*, {0001} in *HCP*, and {111} in *diamond structures*. The facets of cleavage surface are commonly parallel to the cleavage plane.

climb. Displacement of a portion of an *edge dislocation* onto a parallel *slip plane* caused by the contraction (or extension) of its extra-plane, owing to *vacancy* (or *self-interstitial*) transport to the dislocation line.

close-packed direction/row. Straight line in a *crystal lattice,* along which rigid spheres of equal radii representing the atoms are in contact. These directions are: $\langle 110 \rangle$ in *FCC* structure, $\langle 111 \rangle$ in *BCC*, and $\langle 11\bar{2}0 \rangle$ in *HCP.*

close-packed plane. *Lattice plane* with the maximum *lattice point* density: {111} in *FCC* lattice, {110} in *BCC*, and {0001} in *HCP.* They are characterized by the maximum *interplanar spacing*.

closing domain. *Magnetic domain* preventing the appearance of magnetic poles at *high-angle grain boundaries* or at the free surface in the areas where the main domains with antiparallel magnetization vectors touch the interfaces. Closing domains close the magnetic flux inside the corresponding *grain*, thereby reducing the magnetostatic energy (see *domain structure*).

coagulation. Sticking together of particles without their merging. Compare with *coalescence*.

coalescence. Merging of adjacent particles of the same phase, i.e., *subgrains, grains, precipitates, pores,* etc. In the case of *crystalline* particles, coalescence must be accompanied by the disappearance of their *interface* and the simultaneous rotation of at least one of the particles. With regard to precipitates and pores coalescence is sometimes used erroneously instead of *coarsening*.

coarse-grained [Material] characterized by a *mean grain size* greater than 50–100 μm.

coarsening *Thermally activated* process resulting in an increase of the *mean size* of *matrix grains* (*grain coarsening* or *grain growth*) or *precipitates* and *pores* (see *Ostwald ripening*). The *driving force* for coarsening is a decrease of the total *interfacial energy* per unit volume.

coarse pearlite Pearlite formed at the upper limit of *pearlitic range*, i.e., at ~700°C.

Coble creep *Diffusional steady-state creep* whose *strain rate*, $\dot{\varepsilon}$, is controlled by grain-boundary diffusion:

$$\dot{\varepsilon} = a(\sigma/RT)\delta D_{gb}/\overline{D}^3$$

where σ is the *tensile stress*, D_{gb} is the coefficient of grain-boundary self-diffusion, δ is the thickness of *general grain boundary*, $\overline{D}$ is the mean grain diameter, R and T are the gas constant and the absolute temperature, respectively, and a is a coefficient, depending on the grain shape:

$$a = 2/(1 + A + A^2)$$

where A is the *grain aspect ratio* measured in the direction of the applied force.

coherency strain *Elastic deformation* owing to changes in *lattice constants*, e.g., in a diffusion zone. Coherency strains compensate the *lattice misfit* at *coherent interfaces*, e.g., around *GP zones*. Coherency strains are also known as accommodation strains.

coherency strain hardening Increment in *flow stress* due to *dislocation* bending in the field of *coherency strains* around *coherent precipitates*. It is proportional to $f^{1/2}$ (*f* is the volume fraction of the precipitates) and dependent on the magnitude of coherency strains and on the precipitate size.

coherent interface *Phase boundary* wherein the atomic positions in adjoining planes of different *crystal lattices* coincide perfectly or almost perfectly (see Figure C.3). In the latter case, the coincidence is achieved by *coher-*

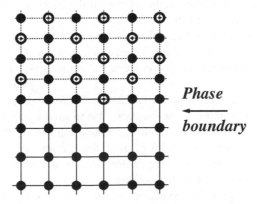

FIGURE C.3 Coherent phase boundary between a solid solution (below) and an intermediate phase. Open and solid circles represent atoms of different components.

ency strains in one of the crystal lattices or both, the *elastic strain energy* being dependent upon the *lattice misfit* and the *elastic moduli* of the crystals. Owing to its *atomic structure*, coherent interface is characterized by the lowest energy among all the types of phase boundaries. It is usually oriented parallel to *close-packed* lattice *planes*.

coherent precipitate *Second-phase* particle with a *coherent interface*. The *lattice* of these precipitates is always specifically oriented relative to the lattice of the *matrix phase* (see *orientation relationship*). For the shape of coherent precipitates, see *nucleus*.

coherent scattering See *elastic scattering*.

coherent twin boundary Interface between the *twin* parts that coincides with the plane of the perfect joining of their *lattices*. Coherent twin boundary, being a *special high-angle boundary* with $\Sigma = 3$ (see *CSL-boundary*), is characterized by low energy and a mobility significantly lower than those of *general grain boundaries*. Under an *optical microscope*, coherent twin boundary looks like a thin, straight line.

coincidence site lattice (CSL) Geometric construction used for the description of the *atomic structure* of *high-angle grain boundaries*. CSL can be constructed as follows. Take two perfectly coincident *crystal lattices*, pass a rotation axis through one of the *lattice sites*, and rotate one lattice around the axis until some (but not all) of its sites coincide with some sites of the other lattice. These coincident sites form a *sublattice* common for both crystal lattices, which is called CSL. *Disorientations* corresponding to different CSL are characterized by Σ values equal to the ratio of the CSL *unit cell volume* to that of the crystal lattice. The same approach can be applied to *low-angle boundaries*; in this case, $\Sigma = 1$. See *special* and *general grain boundaries*.

cold deformation Procedure of *plastic deformation* well below the *recrystallization temperature*.

cold worked Subjected to *cold deformation*.

colony Equiaxed complex formed by two interpenetrating *single crystals* of different *phases*. The crystals appear on the plane sections as alternate lamellae perpendicular to the colony *interface,* or, if the volume fraction of one of the phases in the colony is low, as rod-like branches. Colonies grow into the parent phase by *coupled growth*. Colonies can form because of an *eutectoid* (or eutectic) *decomposition* or *discontinuous precipitation*.

color center *Lattice defect* in optically transparent *ionic crystals* that leads to their coloring. Color centers can occur due to *vacancy* aggregates trapping electrons, which helps to retain the charge balance of the crystals (see, e.g., *F-center, M-center,* and *R-center*). Some *aliovalent solute* (*impurity*) *atoms* can also provide color centers due to disturbances in the *band structure* associated with the solutes.

color etching Techniques for revealing *microstructure* by producing colored films on the surface of different *grains* on polished *metallographic samples*. See, e.g., *staining*.

columnar crystal/grain *Crystallite* of a nearly cylindrical shape. See *columnar structure* and *columnar zone*.

columnar structure *Microstructure* formed by *columnar grains* as, e.g., after *directional solidification* or *zone annealing*; the *grain aspect ratio* in this case can be greater than 10. The columnar structure in thin films consists of grains traversing the film thickness; the grain aspect ratio in this case is ~1. Columnar structure is sometimes observed in a diffusion zone where columnar grains grow parallel to the direction of diffusion flux.

columnar zone Typical feature of *ingot macrostructure*; it is formed by *dendrites* growing in nearly the same direction, tightly to each other. Owing to the anisotropy of dendritic growth (e.g., in metals with *cubic crystal lattices*, the dendrite arms grow along $\langle 001 \rangle$), columnar zone is characterized by a *fiber texture* with a definite axis along the growth direction. Columnar zone is also referred to as transcrystallization zone.

compacted graphite See *vermicular graphite*.

compatibility diagram See *equivalence diagram*.

compensating eyepiece Ocular used in *optical microscopes* in combination with *apochromatic objectives*.

complex carbide In *steels*, *carbide* containing, along with iron, other metallic *components*, e.g., $(Cr,Fe)_{23}C_6$.

component In materials science, a substance, usually a chemical element or a compound of a constant *composition*, forming different *phases* in a *system*. Components can exist independently of the system.

composite Material designed of chemically different and insoluble constituents. Commonly, these constituents are macroscopic, but in some cases, they can be microscopic as, e.g., in *ODS alloys* and *ZTA*.

composition Number of elementary constituents: *components* in an *alloy* (*phase*), e.g., single-component, *binary, ternary*, multicomponent, etc., or phases in an alloy, e.g., *single-phase (homogeneous), multiphase (heterophase)*, etc. The same term also denotes the content of components. See also *phase composition* and *concentration*.

compromise texture *Annealing texture* whose formation can be explained by approximately equal *mobility* of the boundaries between the growing *grains* and different components of the matrix texture.

Compton scattering *Inelastic* x-ray *scattering* on free or weakly bounded electrons in the substance. This scattering mode contributes to the *background* in x-ray diffraction patterns.

concentration Relative content of a *component* in an *alloy* (*phase*) expressed in *at%, mol%*, or *wt%*. The same term can denote the relative number of *point defects* (see, e.g., *vacancy*).

condensed atmosphere See *Cottrell atmosphere*.

conduction band See *band structure*.

congruent [*Phase transition*, commonly *melting*] that evolves without any compositional alterations in a participating *solid* phase. In contrast, melting accompanied by decomposition of a solid phase, as, e.g., in *peritectic reaction*, is termed incongruent.

conjugate slip system *Slip system* becoming *active* after the *resolved shear stress* on a *primary* slip system decreases below *critical* value, due to a lattice rotation in the course of *plastic deformation*. See *Schmid's law*.

conode See *tie line*.

constitution diagram See *phase diagram*.

constitutional undercooling/supercooling *Undercooling* of melt due to changes in its equilibrium *solidification temperature*, T_0, caused by changes in the *solute concentration* next to the solid/liquid *interface*. If the actual melt temperature is lower than T_0, the melt is constitutionally *undercooled*, which results in an instability of the planar shape of the interface. This instability leads to the evolution of *cellular* or *dendritic structures,* depending on the undercooling and the *linear growth rate*.

constraint Restriction of a process, e.g., of the crystal growth or plastic deformation.

continuous cooling transformation (CCT) diagram Presentation of the evolution of *phase transformations* at different cooling rates by lines, in coordinates temperature–time, corresponding to the transformation start and its finish.

continuous grain growth See *normal grain growth*.

continuous precipitation *Phase transformation* evolving according to the occurrence and *diffusion-controlled* growth of new phase *precipitates* inside the parent phase.

continuous recrystallization Process of *microstructural* alterations on *annealing plastically deformed* material, which results in a decrease of the overall *dislocation density* and the formation of *strain*-free *subgrains*. A further annealing is accompanied by subgrain growth only, identical to *normal grain growth*. This can be explained by the exhaustion of the *elastic strain energy* associated with *lattice defects*, which might contribute to the formation of *recrystallization nuclei*. Continuous recrystallization is observed in materials of high *stacking-fault energy* (e.g., in Al alloys), where dislocation rearrangements, leading to the formation of *subgrain structure*, evolve fast. *Precipitation* from *supersaturated solid solution* enhances continuous recrystallization because it retards the evolution of recrystallization nuclei. The latter is a result of the preferential arrangement of precipitates at *subboundaries* (see *heterogeneous nucleation*), which inhibits their migration (see *particle drag*). Because of this, an increase of the *disorientation* angle at the subboundaries, their transformation into *high-angle boundaries*, and the formation of the recrystallization nuclei are inhibited. Continuous recrystallization is also called recrystallization *in situ*.

continuous [x-ray] spectrum See *white radiation*.

controlled rolling *Thermo-mechanical treatment* of low-*alloy steels* with <0.1 wt% C, aiming at increasing their toughness and strength by decreasing *ferrite* grain size (see *grain-boundary strengthening*), as well as by increasing the volume fraction of *dispersed phases* (see *precipitation strengthening*). The main stage of controlled rolling is *hot deformation*, in the course of which the temperature reduces from the γ-field to the (γ + α)-field in Fe–Fe₃C diagram. *Dynamic recrystallization* of *austenite* during the hot deformation results in a decrease of its grain size, whereas *grain growth* is inhibited by *dispersed particles* of *alloy carbides, nitrides*, or *carbonitrides* precipitating at the austenite *grain boundaries*. The second main stage is a rapid cooling of the hot-deformed article, in which ferrite occurs from austenite. At this stage, *metadynamic recrystallization* of austenite is inhibited, which retains small austenite grain size and leads to a decreased (5–10 μm) ferrite grain size.

convergent beam electron diffraction (CBED) Technique used in *TEM* for determining both the *crystal lattice* and its orientation in areas of linear size ~5 nm.

cooperative growth See *coupled growth*.

coordination number (CN) Number of the first nearest neighbors of any atom or ion in a *crystal structure*. It equals 12 in *FCC* and *HCP*, 8 in *BCC*, and 4 in *diamond* structures.

coordination polyhedron Polyhedron whose vertices lie at the centers of the atoms (ions) of the first *coordination shell*. In *covalent crystals*, a coordination polyhedron has a shape determined by the number of valence electrons forming covalent bonds, whereas in *metallic* and *ionic crystals*, its shape is governed by the ratio of the atom (ion) sizes, the ratio being known as a geometric factor. In the case that atoms (ions) are taken to be rigid spheres, some of them being of a small radius r and others of a greater radius R, and small atoms (ions) are surrounded by greater ones and vice versa, the shape of coordination polyhedrons is determined by the ratio r/R. For instance, if r/R is between 0.225 and 0.414, the greater atoms (ions) form a tetrahedron with the *coordination number* CN = 4; if $0.414 < r/R ≤ 0.732$, an octahedron (CN = 6); if $0.732 < r/R ≤ 1.0$, a cube (CN = 8). If r/R corresponds to the upper limit of the ranges named, all the atoms (ions) are in contact, and the corresponding polyhedron is regular.

coordination shell Shell of an atom or ion formed by the nearest neighbors in a *crystal structure*; it is called the first coordination shell. The second nearest neighbors form the second coordination shell, and so forth.

core segregation See *coring*.

coring *Microsegregation* in *dendrites* of *phases* that change chemical *composition* in the course of *solidification*. It reveals itself in the variation of *solute concentration* from the core to the periphery of the dendritic arm. The dendrite core is enriched by a *component* increasing the *solidification temperature* in the concentration range concerned, which depends on the configuration of the *liquidus* line and not just on the solidification temperature of the components. Microscopic *chemical inhomogeneity* occurs due to a limited *diffusivity* in the solid state. In an alloy X (see Figure C.4), the grain cores of α-phase are enriched by atoms A, although the component A has a lower *solidification point* than B. Coring is also known as dendritic (or core) segregation.

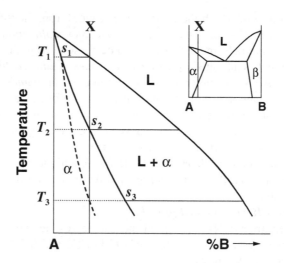

FIGURE C.4 Coring formation in α solid solution in the course of its solidification. Points s_1 and s_2 show compositions of α-crystallites occurring at temperatures T_1 and T_2. According to the phase diagram, α-phase at T_2 should be of composition s_2. However, owing to low diffusivity in the solid state, the attainment of the equilibrium composition evolves slowly. Thus, it cannot be reached at an increased cooling rate. An averaged composition of α-phase, displayed by a dashed line, shows that the solidification of α-phase can be completed only at temperature T_3. Thus, the composition of α-crystallites will vary, from s_1 in their core, to s_3 at the periphery.

corundum *Polycrystalline* aluminium oxide Al_2O_3.

Cottrell atmosphere *Segregation* of *interstitial solutes* at *edge dislocations* accompanied by a decrease of the overall *elastic strain energy* due to the fact that the *lattice distortions* associated with the solute atoms are partially compensated by the lattice distortions associated with the dislocations. Cottrell atmosphere is called condensed if the solute atoms form a continuous string along the dislocation line. The solute *concentration* necessary to produce condensed atmospheres at all the dislocations is $\sim 10^{-4}$ at% at the *dislocation density* $\sim 10^{10}$ cm^{-2}.

Cottrell cloud See *Cottrell atmosphere.*

coupled growth Growth of contacting grains of different *phases*, which results in the formation of a *colony*. Such a simultaneous growth reduces the total energy of *interfaces* between the growing grains and the parent phase because the grains only partially border on the parent phase, while the interface between the grains is commonly of low energy. Since the phases in the colony are always of different chemical *compositions*, their growth into the parent phase should be accompanied by the diffusional redistribution of the *alloying elements*. Thus, coupled growth is a *diffusion-controlled* reaction via diffusion, either through the parent phase ahead of the colony, or over the colony interface. Colonies always grow faster than separated grains of the same phases, because the diffusion paths for the growth of crystallites in the colony are relatively short, of the order of an *interlamellar spacing*, whereas much longer diffusion paths are necessary for the growth of separated grains. Coupled growth is also referred to as cooperative growth.

covalent bond Interatomic bond resulting from the formation of common pairs of valence electrons with the opposite spins. This bond is strictly oriented in space. It is also referred to as homopolar bond.

covalent crystal Crystal whose atoms are connected by *covalent bonds*; in some cases, the bond may be partially *ionic* (see *electronegativity*). A *coordination number* in covalent crystals equals $(8 - N)$, where N is the number of valence electrons. Owing to high anisotropy of the bond, covalent crystals have loosely packed *lattices* (see, e.g., *diamond structure* and *zinc blende structure*). *Lattice defects* disturb the *band structure* and can act as *shallow impurities* and *deep centers*.

covalent radius *Atomic radius* in *covalent crystals* defined as half the shortest distance between identical nonmetallic atoms. An octahedral covalent radius (in crystals with *coordination number* equals 6) and a tetrahedral covalent radius (coordination number equals 4) are to be discerned.

creep *Plastic deformation* evolving with time at a constant temperature (above ~0.3 T_m) and a constant *stress* lower than the *yield stress* at room temperature. At temperatures <0.4 T_m and low stresses, *logarithmic creep* is observed. At temperatures >0.4 T_m, three creep stages are observed: the first one with a decreasing deformation rate (primary, or transient, creep), the second one with a constant deformation rate (secondary, or steady-state, creep), and the third one with an increasing deformation rate (tertiary creep). There are several creep mechanisms: *dislocation glide* motion, at $\sigma/E > 10^{-2}$; *dislocation creep*, at $\sigma/E = 10^{-2}$–10^{-4}; and *diffusional creep*, at $\sigma/E < 10^{-5}$ (σ and E are the *tensile stress* and *Young's modulus*, respectively).

creep cavitation Occurrence of voids at *grain boundaries* arranged perpendicular to the direction of *tensile stresses*. The voids form mainly due to *diffusional plasticity*, especially at the boundaries where *grain-boundary sliding* takes place or where *precipitate* particles are arranged. The voids coarsen with time due to vacancy flux between the voids of different sizes (see *Ostwald ripening*), or unite in cracks, which eventually leads to the creep rupture. Creep cavitation can noticeably contribute to the creep deformation in ceramic materials.

cristobalite High-temperature *polymorphic modification* of *silica* known as "high," or β-, cristobalite. It has a *cubic crystal structure* and, on cooling at an increased rate, transforms at ~250°C into a *metastable* "low," or α-, cristobalite with *tetragonal* structure, the transformation being *displacive*.

critical cooling rate Minimum rate of continuous cooling of a high-temperature *phase* at which no *diffusional transformation* develops. Cooling at a greater rate results in the retention of the high-temperature phase or in the occurrence of a *metastable* phase. For instance, a *supersaturated solid solution* can be obtained by cooling a *solid solution* at a critical rate, which prevents *precipitation* of a new phase according to the *phase diagram*. A metastable *glassy phase* can be obtained from a liquid phase by cooling the melt at some critical rate, which prevents the occurrence of more stable *crystalline* phases. In the case that a high-temperature solid phase undergoes *polymorphic transformation*, cooling at a critical rate can result in *martensitic transformation*, as, e.g., in *steels* and Ti *alloys*. The magnitude of critical cooling rate depends on the *composition* of the corresponding high-temperature phase.

critical deformation Magnitude of *plastic deformation* necessary for *recrystallization nuclei* to occur at subsequent *annealing* (i.e., at *static recrystallization*) or in the course of deformation (i.e., at *dynamic recrystallization*). In the case of *cold deformation*, it corresponds to a sharp peak on the dependence of the grain size after annealing versus deformation degree (see *recrystallization diagram*), and does not exceed 5–10%, depending on the material purity and the initial grain size. The annealing after critical deformation (CD) results in *abnormal grain growth* when large grains with a decreased *dislocation density* grow (see *strain-induced grain boundary migration*). The grain size corresponding to CD is up to 20–30 times greater than at higher deformations. This is used for receiving large (up to several cm) metallic *single crystals*. In the case of *hot deformation*, CD corresponds to a *strain* value at which dynamic recrystallization starts. The magnitude of CD in this case depends on the deformation rate and temperature.

critical point In material science, an actual temperature at which a certain *phase transformation* commences. It can be either close to the *equilibrium temperature*, T_0 in the corresponding *phase diagram*, or much lower than T_0, which depends on the cooling (heating) rate (see *undercooling* and *superheating*). In *thermal analysis*, critical points are termed arrest points.

critical-resolved shear stress Shear stress necessary for initiating the *dislocation glide* over a *slip system*. In *strain*-free materials, it usually equals the *Peierls stress*.

critical [size] nucleus *Crystallite* of a new phase that can become either stable on addition of one atom or unstable on removal of one atom. Its size can be estimated as:

$$r_{cr} = k\sigma/(\Delta g_{tr} - g_\varepsilon)$$

where k is a coefficient depending on the nucleus shape, σ is its *interfacial energy*, Δg_{tr} is the driving force (per unit volume) for the *phase transition* considered, and g_ε is the specific (per unit volume) *elastic strain energy* associated with the change in the *specific volume*. The *free energy*, Δg, necessary for critical nucleus to occur is:

$$\Delta g = k'\sigma^3/(\Delta g_{tr} - g_\varepsilon)^2$$

where k' is a coefficient depending on the nucleus shape. This energy is required for creating a new *interface* between the nucleus and the parent phase, as well as for compensating the above-mentioned *elastic strain energy*. Thus, the nucleation of a new phase requires *fluctuations* whose magnitude is not smaller than Δg_{cr}. Since Δg_{tr} is approximately proportional to *undercooling*, ΔT, both of the formulae show that critical nuclei occur at $\Delta T > 0$ and cannot occur at $\Delta T = 0$. As follows from the previously described equations, Δg_{cr} decreases with the reduction of the interface energy σ. Thus, small crystals of a third phase, present in the matrix before the nucleation starts, and having a decreased σ with the new phase, facilitate nucleation of the latter. This is used for decreasing the grain size of new phase (see *inoculant*, *nucleation agent*, and *glass-ceramic*). See also *heterogeneous nucleation*.

cross-slip *Glide* motion of a *screw dislocation* passing from the *primary slip plane* onto another, nonparallel slip plane. The intersection line of the planes must be parallel to the *Burgers vector* of the dislocation.

crowdion Portion of a *close-packed* atomic row with a decreased *interatomic spacing* due to an extra atom. Crowdions occur under irradiation by high-energy ions or thermal neutrons.

crystal *Solid* body characterized by a *long-range order* in its *atomic structure* and by a regular shape with flat facets and definite angles between them, the angles being dependent upon its *lattice*.

crystal axis Vector coinciding with an edge of a *unit cell*, its length being equal to the edge length. It is also known as *fundamental translation vector*.

crystal defect Disturbance of a periodic atomic arrangement in a *crystal lattice*. *Point defects*, *linear defects*, and *planar defects* are usually considered.

crystal imperfection See *crystal defect*.

crystal lattice Repeating three-dimensional pattern of *lattice points*, each point having the same surrounding. Crystal lattice can be obtained by the translation of a *unit cell* along the *crystal axes*. Because of this, any crystal lattice possesses a translational microscopic symmetry. Crystal lattice is also referred to as point lattice or simply lattice.

crystalline Having a certain *crystal structure*, i.e., characterized by both the *short-range* and *long-range order* in the *atomic structure*.

crystalline anisotropy Orientation dependence of various properties in a *crystal lattice*.

crystalline ceramic Inorganic, nonmetallic material obtained mostly by a high-temperature treatment (known as *firing*) of particulate products, causing *sintering* and other solid-state reactions. In some cases, ceramics can be obtained by *crystallization* of a *glassy phase* (see *glass-ceramic*), vapor deposition, etc.

crystalline fracture Brittle fracture wherein an increased area of the fracture surface lies parallel to *cleavage planes*.

crystallite See *grain*.

crystallization Formation of *crystalline phases* upon cooling a *liquid* phase (see *solidification*) or upon heating an *amorphous* phase; the latter is sometimes referred to as *recrystallization*. The formation of crystalline phases upon heating a *glassy* phase is called *devitrification*, but is also known as crystallization.

crystallization point/temperature See *solidification*.

crystallographic texture Preferred orientation of *crystal lattices* of the majority of *grains* with respect to some coordinate system. Usually, the system is related to the specimen shape or its production scheme. For instance, in wires, the principal axis of the coordinate system is the wire axis, whereas in sheets, the principal axes are *ND, RD, TD*, etc. Texture is qualitatively characterized by the number and type of *texture components*, and quantitatively by their *intensity* and *scatter*. If texture has only one component, it is called single-component, and in the opposite case, multicomponent. Textures can evolve during *solidification* (especially during *directional solidification*); electrolytic, vacuum or sputter deposition; *plastic deformation; recrystallization; phase transitions;* etc. Polycrystalline *textured* materials are always *anisotropic*. In materials science, the term texture is frequently used instead of crystallographic texture.

crystal monochromator Flat or slightly bent *single crystal* placed between the x-ray source and the specimen. The x-ray beam diffracted on the crystal is more monochromatic than the primary beam.

crystal structure Atomic arrangement that can be received by setting atoms (or ions or atom groups) in connection with the *lattice points* of a *crystal lattice*.

crystal system Description of a *crystal lattice* according to the shape of its *unit cell: cubic, tetragonal, orthorhombic* (or *rhombic*), *rhombohedral* (or *trigonal*), *hexagonal, monoclinic*, and *triclinic*. Each crystal system must possess definite macroscopic *symmetry elements*: triclinic, none; monoclinic, a single 2-fold *symmetry axis* or a single *mirror plane;* orthorhombic, three mutually perpendicular 2-fold axes or two perpendicular mirror planes; tetragonal, a single 4-fold axis or a rotation-inversion axis; rhombohedral, a single 3-fold axis or a rotation-inversion axis; hexagonal, a single 6-fold axis or a rotation-inversion axis; and cubic, four 3-fold axes.

CsCl structure [type] *Crystal structure* wherein cation and anion *sublattices* are *cubic primitive* and shifted in respect to one another by $1/2\ a\langle 111\rangle$, where a is the *lattice constant*. CsCl structure is analogous to *BCC structure* (see Figure C.5), except it is formed by atoms of two kinds. CsCl structure type is characteristic of *ionic crystals*.

CSL-boundary See *special grain boundary*.

cube-on-edge texture See *Goss texture*.

cube orientation {100}[001] *component* observed in the *annealing texture* of *FCC* metallic materials with high *stacking-fault energy* as, e.g., in Al-based *alloys*.

cube texture Single-component *annealing texture* characterized by alignment of *lattice planes* {100} parallel to the tape (strip) plane and $\langle 001\rangle$ directions

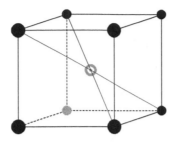

FIGURE C.5 Unit cell of crystal structure CsCl; solid and open spheres denote various atoms.

parallel to the rolling direction. It is usually observed in thin tapes of *FCC* metallic materials with medium *stacking-fault energy*. The preceding *cold-rolling* texture should have the following main components: {112}⟨111⟩, {011}⟨211⟩ and {123}⟨634⟩. The *intensity* of cube texture can be quite high and its *scatter* quite low (<2°). In some soft-magnetic Fe–Ni alloys, cube texture improves magnetic properties because ⟨001⟩ in these alloys is an *easy magnetization direction* (see also *magnetic texture*). Cube texture, and even a noticeable *cube orientation,* is detrimental in sheets for deep drawing, because it increases *planar anisotropy.*

cubic martensite In *steels*, martensite with *tetragonality* close to 1.0. This can be a result of a low carbon content in *solid solution* (see *tempered martensite*).

cubic system *Crystal system* whose *unit cell* is characterized by the following *lattice parameters*: $a = b = c$, $\alpha = \beta = \gamma = 90°$.

Curie/temperature point (T_c, Θ_c) In materials with *magnetic ordering*, the temperature of the transition of *ferromagnetic (ferrimagnetic) phase* into *paramagnetic* phase, or vice versa. In *ferroelectrics*, a temperature of *polymorphic transformation* cubic ↔ non-cubic, resulting in the appearance or disappearance of electrical polarization, is also called Curie point.

curvature-driven grain growth *Grain growth* under the influence of the *capillary driving force.*

Cu-type orientation One of the main *deformation texture components*, {112}⟨111⟩, observed in *cold-rolled FCC* metallic materials of medium to high *stacking-fault energy.*

D

δ-Fe *Allotropic form* of iron having *BCC crystal structure* and existing at atmospheric pressure at temperatures above A_4 (i.e., > 1400°C) up to the *melting point*.

δ-ferrite *Solid solution* of *alloying elements* and/or carbon in δ-Fe.

Δr-value Quantity characterizing *planar anisotropy* in sheets:

$$\Delta r = (r_0 + r_{90} - 2r_{45})/4$$

where r is the *r-value*, and the subscripts 0, 45, and 90 denote the angles between the axis of the *tension* specimen and *RD* of the sheet.

dark-field illumination In *optical microscopes*, such illumination that flat horizontal features of an opaque sample appear dark, whereas all the inclined features appear bright (e.g., in single-phase materials, *grains* are dark and *grain boundaries* bright, in contrast to *bright-field illumination*). This is due to the fact that the inclined features reflect the incident light into an objective, whereas the horizontal features do not.

dark-field image High-resolution image produced by a diffracted beam directed along the *TEM* axis. The contrast in a dark-field image is opposite to that in a *bright-field* one, e.g., a *dislocation* line is bright in the former and dark in the latter.

Debye–Scherrer method *Powder method* wherein a needle-like *polycrystalline* specimen is placed along the axis of a cylindrical camera and a film is placed inside the camera on its wall. A monochromatic and collimated primary x-ray beam is directed onto the specimen along the camera diameter. The specimen can be rotated during the exposition.

decomposition In a *phase diagram* with a *miscibility gap*, a *phase transition* in a *solid solution*, α, which decomposes upon cooling into *isomorphous solid solutions*, α_1 and α_2 (see Figure D.1). The same term is used to designate certain phase transformations in the solid state, e.g., *eutectoid decomposition*, decomposition of *supersaturated solid solution* on *aging treatment* into *precipitates* and *saturated solid solution*, decomposition of *cementite* on *graphitization* into *austenite* and *graphite*, etc. In the two latter cases, decomposition is connected with the transformation of a *metastable phase* into more *stable* ones.

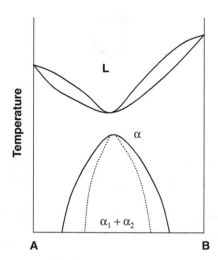

FIGURE D.1 Binary diagram with a miscibility gap in the solid state; α_1 and α_2 denote α solid solutions of different compositions.

decorated dislocation Dislocation with *precipitate particles* along its line. The particles occur due to *equilibrium segregation* of *interstitial solutes* (see *Cottrell atmosphere*) and nucleate at the dislocation (see *heterogeneous nucleation*). The particles inhibit the dislocation *glide*, which manifests itself in the appearance of *sharp yield point, strain aging*, or *dynamic strain aging*.

deep center *Lattice defect* in semiconductors and insulators, whose energy level lies inside the *band gap* and is characterized by an activation energy higher than kT (k is the *Boltzmann constant,* and T is the absolute temperature). Deep centers act as carrier traps and recombination centers for charge carriers of opposite signs.

defect structure/lattice *Crystal structure* of an *intermediate phase* with some degree of *structural disorder.*

deformation band Elongated area of noticeable thickness (several μm) inside a deformed *crystallite*. The area rotates in the course of *plastic deformation* (see *slip*) differently than the rest of the crystallite. As a result, an overall disorientation as high as 50–70° can develop across the deformation band. Deformation bands are formed by up to 15 of parallel rows of elongated dislocation cells (see *cell structure*). The disorientation between the neighboring rows is several degrees. Deformation band is also known as transition band or microband.

deformation kinking Occurrence of *kink bands* during *plastic deformation*. It is observed in compression tests of *single crystals* having a small number of possible *slip systems*, e.g., in *HCP* crystals.

deformation mechanism *Plastic deformation* in *crystalline* solids can develop via different atomic mechanisms: *slip, deformation twinning, grain-boundary sliding,* and *diffusional plasticity.* Slip and deformation twinning are competing mechanisms. The latter mostly operates as long as slip is inhibited, e.g., at high strain rates, at decreased temperatures, or at a small number of *slip systems* (as, e.g., in *HCP* metals). The latter two mechanisms operate at temperatures $>0.5\ T_m$ and low *strain rates* only.

deformation mechanism map Graphic representation of the main *deformation mechanisms* in coordinates σ/E – *homologous temperature* (see Figure D.2), or sometimes $\sigma/E - \overline{D}/b$, where σ is the *tensile stress*, E is the *Young modulus*, $\overline{D}$ is the *mean grain size*, and b is the *Burgers vector*.

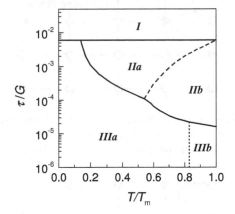

FIGURE D.2 Deformation mechanism map for coarse-grained Ni (scheme): (I) slip, (II) dislocation creep controlled by (IIa) cross-slip or (IIb) climb, (III) diffusional creep controlled by (IIIa) grain boundary diffusion or (IIIb) bulk diffusion.

deformation texture *Preferred grain orientation* forming in the course of *plastic deformation*. The character, *intensity*, and *scatter* of deformation texture depend on the loading scheme, deformation degree, and temperature, as well as on the initial *microstructure* and texture. Deformation texture develops because the grain *lattices* rotate during deformation in such a way that *active slip systems* in different grains tend to be directed along the principal *tension* deformations of the sample.

deformation twin Twin occurring in the course of *plastic deformation*. Under an *optical microscope*, deformation twins look like thin lamellae that never intersect the *grain boundaries*. See also *Neumann band* and *deformation twinning*.

deformation twinning Mechanism of *plastic deformation* revealing itself in a discrete change of the orientation of a *crystal lattice* into that of a *twin*. It is typical of *BCC*, *HCP*, and some *FCC* metallic *alloys*, and is favored by both a decrease in the deformation temperature and an increase in the *strain rate*. Deformation twinning (DT) can be described as a *shear* over *twinning system*, the shear magnitude being proportional to the distance from the twinning plane (see Figure D.3). In this respect, DT differs from *slip* (see Figure D.4). Another characteristic feature of DT is a high rate of the twin development, much higher than the rate of the dislocation

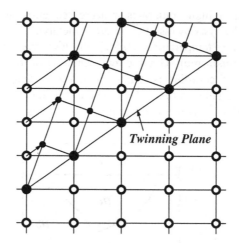

FIGURE D.3 Deformation twinning. Open circles show atomic positions before deformation; solid circles correspond to the positions after deformation. Arrows show atomic displacements, different for various atomic layers parallel to the twinning plane.

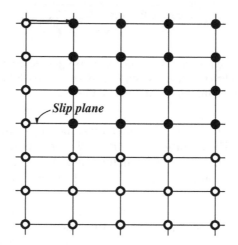

FIGURE D.4 Slip deformation. Open circles show atomic positions before deformation; solid circles correspond to the positions after deformation. Atomic displacements in the deformed part are the same for various atomic layers parallel to the slip plane (see arrow).

glide motion in slip. DT is supposed to develop owing to the glide of *Shockley partial dislocations* initiated by a certain *resolved shear stress*. To produce a twin, the twinning dislocation should glide subsequently over all the twinning planes forming the twin body. The maximum *true strain* resulting from DT can reach 0.4.

degree of freedom In thermodynamics and materials science, a maximum number of variables that can be changed independently without changing the *phase composition* of an isolated *system* at *thermodynamic equilibrium*. For instance, in a *single-component* system, these variables are temperature and pressure. Possible degrees of freedom in *binary* systems are temperature, pressure, and the *concentration* of one of the *components* in one of the *phases* (the concentration of the other component cannot be varied arbitrarily). In *ternary* systems, the possible degrees of freedom are temperature, pressure, and concentrations of two components. See *Gibbs' phase rule* and *tie line*.

degree of long-range order See *long-range order parameter*.

dendrite *Crystallite* of a tree-like shape. Such a shape results from the growth into the melt (or *amorphous phase*) under significant *undercooling*, either thermal or *constitutional*. If the growth is unconstrained, the arms of a higher order develop from the main dendrite arm (called primary arm). *Lattice directions* of all the dendrite arms belong to the same *form*. If the primary arms of the neighboring dendrites are close to each other, a *columnar zone* develops.

dendritic segregation See *coring*.

densely packed plane *Lattice plane* with an increased density of *lattice points*; it is characterized by an increased *interplanar spacing* and low *Miller indices*. The most densely packed plane is called the *close-packed plane*.

depth of focus Range along the light (electron) beam axis wherein all the points on the object's surface are focused simultaneously. A high depth of focus makes it possible to focus all the peaks and valleys on a rough sample surface.

desorption Process opposite to *adsorption*. It results in the disappearance of an adsorption layer, e.g., due to an increase in temperature or a decrease of the *adsorbate concentration* in the environment.

devitrification *Crystallization* of a *glassy phase*; it can evolve due to the appearance of either one or many *crystalline phases*. In the former case, the crystalline phase has the same chemical *composition* as the original glassy phase (a process known as polymorphic crystallization); in the latter case, compositions of the crystalline phases differ from that of the glassy phase, as in *eutectic reactions*. The *nucleation* of the crystalline phases is mostly *homogeneous*, but can also be *heterogeneous*, owing to *nucleation agents* or quenched-in nuclei (small *crystals* that may be present in the glassy phase). In some cases, devitrification can be preceded by the occurrence of chemically different amorphous phases and a decrease of the *excess volume*.

diamagnetic Material having no magnetic moment of its own. It can be slightly magnetized by an external magnetic field in the direction opposite to the field direction.

diamond structure *Crystal structure* in materials with a *covalent interatomic bond* wherein each atom has four nearest neighbors in the vertices of a regular tetrahedron (see Figure D.5). The structure is typical of C (a diamond *allotropic modification*), Si, and Ge. It is analogous to the *zinc blende structure* (see Figure Z.1), but in the diamond structure, all the atoms are the same kind.

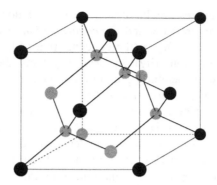

FIGURE D.5 Unit cell of diamond structure.

differential interference contrast Technique of *optical microscopy* used for studying a surface relief with a height difference from ~1 to tens of wavelengths, using the interferometry of polarized light.

differential scanning calorimetry (DSC) Technique for determining the rate of the heat evolution (or *absorption*) by comparing the behavior of a specimen with that of an inert standard, both subjected to a controlled cooling (heating). The data are presented as dependence $\Delta Q/\Delta T$ vs. T, where Q is the released (absorbed) heat, and T is the temperature.

differential thermal analysis (DTA) Technique for determining the temperatures of *phase transitions* during heating (cooling) a specimen and an inert standard by plotting their temperature difference versus temperature. There must be no phase transitions in the standard in the temperature range studied.

diffraction angle Angle 2θ between the reflected and primary beams in the *diffractometric method*.

diffraction contrast In *TEM*, an image contrast produced by electrons diffracted from a *crystal lattice*. It is also known as phase contrast.

diffraction spot *Dot* in x-ray diffraction patterns of *single crystals* or *selected area diffraction patterns*.

diffractogram Pattern recorded in the course of investigations by the *diffractometric method*.

diffractometer Device registering the intensity and direction of x-ray beams diffracted from a specimen (see *diffractometric method*). A diffractometer is commonly used for *x-ray structure analysis*; there are also specialized diffractometers for determining *textures, macroscopic stresses, single-crystal* orientation, etc.

diffractometric method Technique for *x-ray structure analysis*. In this technique, a sample is adjusted in such a way that its surface coincides with the axis of the diffractometer circle, and the sample can be rotated around the axis. The focal spot of the x-ray tube, producing a *characteristic* x-ray beam, as well as the counter window, are both located on the diffractometer circle. The counter is rotated at an angular rate two times greater than that of the sample (i.e., the counter is always at a *diffraction angle*), and the diffracted beam can be registered. The counter-rotation is synchronized with the motion of the strip chart of a recorder. The latter yields a *diffractogram* displaying the intensity of diffracted radiation versus diffraction angle.

diffuse scattering X-ray (electron) scattering in directions remote from those corresponding to the *Bragg condition*. The main sources of diffuse scattering in *crystalline solids* are thermal vibrations of atoms around the *lattice sites* and *static lattice distortions* associated with *solute* atoms and other *crystal defects*. Diffuse scattering is noticeably increased by *amorphous phases* present in the sample studied.

diffusion *Thermally activated* process of mass transport, reducing *concentration* gradients and evolving by a random walk of atoms. The diffusion of atoms of a certain type at a constant temperature is macroscopically described by *Fick's laws*. In case of superposition of a *drift* motion tending to decrease the gradient of *chemical potential*, a concentration gradient can increase (see *uphill diffusion*). Diffusion processes in *crystalline solids* are categorized according to: the type of atoms involved (*self-diffusion* and *solute diffusion*); the diffusion path (surface diffusion, *grain-boundary diffusion, pipe diffusion,* and *bulk diffusion*); specific conditions (e.g., *chemical diffusion, ambipolar diffusion, uphill diffusion, electromigration,* etc.); and the diffusion mechanism (*interstitial diffusion* or *vacancy mechanism*).

diffusional creep High-temperature creep at low *stresses* whose *strain rate* is controlled by mass transport over *grain boundaries* (*Coble creep*) or through the body of *crystallites* (*Nabarro–Herring creep*). If diffusional creep develops in a *dispersion strengthened* material, *precipitation-free zones* occur that delineate the grain boundaries arranging perpendicular to the loading direction. See also *diffusional plasticity*.

diffusional plasticity Mechanism of *plastic deformation* at temperatures >0.5 T_m and low *strain rates*. It evolves because of a directional flux of *vacancies* (or by an atomic flux in the opposite direction, which is the same thing) in the *strain* field induced by external force. For instance, a vacancy flux is directed toward the compressed regions from the regions subjected to *tension*. As a result, each grain, and the specimen as a whole, increases its length and decreases its transverse size. The *grain boundaries* in diffusional plasticity serve as the *sources* of the vacancies and as their *sinks*.

diffusional transformation *Phase transition* characterized by the growth of new *grains* wherein individual atoms cross the *interface* between a new phase and the parent one by uncoordinated, *thermally activated* jumps. The new phase can be of the same *composition* as the parent phase or a different one. The main features of this transformation are the following. It evolves with time at a constant temperature (i.e., *isothermally*), and its *kinetics* depend on the *nucleation rate* and the *linear growth rate* of new grains or *colonies* (see *Avrami* and *Johnson–Mehl–Kolmogorov equations*). Both the incubation period and the transformation rate depend on *undercooling*, ΔT. The period is large and the rate low at both a low and large ΔT; in the first case, due to a low *nucleation rate*, and in the second one, due to a low growth rate (see *C-curve*). *Lattice defects,* such as *vacancies*, *dislocations*, and *grain boundaries* increase the nucleation rate, favoring the new phase *nucleation* (see *heterogeneous nucleation*), as well as the linear growth rate at relatively low temperatures, because many lattice defects serve as *short-circuit diffusion paths*. An *orientation relationship* might be observed between the new phase and the parent one, as well as between the new grains in a colony. The linear growth rate and the shape of the new phase grains depends on the *atomic structure* and orientation of their boundaries (see *coherent, partially coherent,* and *incoherent* interfaces). In cases in which the interface is incoherent, the growth is *diffusion-controlled* and the linear growth rate is *isotropic*, because the atomic flux is homogeneous over the interface area. All this results in an *equiaxed* shape of new grains. When a perfectly or partially coherent interface consists of wide, atomically smooth terraces connected by ledges, the interface can migrate because of the lateral motion of the ledges. In this case, the growth is *interface-controlled* and the growth rate is highly *anisotropic*, which results in a plate-like shape of the crystallites. The growth rate can also be interface-controlled if it depends upon the reaction rate at the interface. Diffusional transformation is also termed reconstructive.

diffusion coefficient Coefficient of proportionality, D, in *Fick's laws* (its units are m^2/s). Coefficients of *self-diffusion* and *solute diffusion,* as well as those of *bulk, grain-boundary*, and surface diffusion differ significantly. In *ionic crystals*, diffusion coefficients of cations and anions also differ owing to unequal sizes of these ions. The temperature dependence of diffusion coefficients obeys the *Arrhenius equation*. Diffusion coefficient can be referred to as diffusivity. See also *chemical diffusion*.

diffusion-controlled Dependent upon the *diffusion* rate only.

diffusion-induced grain boundary migration (DIGM) Displacement of a *grain boundary* caused by a flux of *solute* atoms over it. The solutes also diffuse into, or out of, the body of the *grains* separated by the boundary.

diffusion-induced recrystallization (DIR) Occurrence of new *grains* at a *grain boundary,* over which there exists a flux of *solute* atoms.

diffusionless transformation *Phase transition* in which new *grains* grow via the coordinated displacement of atoms, the displacement being smaller than the *interatomic spacing* in a parent phase. The *interface* between the parent and new phases should be *coherent*. The grains of the new phase always have the same *composition* as the grains of the parent phase. Diffusionless transformation is also referred to as *displacive, nondiffusional* or *shear-type transformation*. See, e.g., *martensitic transformation* or the formation of ω-*phase* in Ti alloys.

diffusion porosity See *Kirkendall effect*.

diffusivity See *diffusion coefficient*.

dilation Relative volume alteration of a solid body, usually under the influence of hydrostatic pressure, or owing to thermal expansion or *phase transfor-*

mations. If the properties of the body are *isotropic*, dilation does not lead to changes in its shape.

dilatometer Device used for measuring linear size alterations of a specimen. The alterations may be caused by thermal expansion or *phase transitions* involving volume changes, the latter resulting from the difference in *specific volume* between the parent and the new phases.

dilute [solid] solution *Solid solution* with a *solute* content well below its *solubility limit*.

direct replica See *replica*.

directional solidification Procedure for producing *castings* with a *macrostructure* characterized by a *columnar zone* only. This can be achieved by directing the temperature gradient in the melt along the axis of the growing *columnar grains*. *Single crystals* can also be obtained in such a way.

disclination *Linear lattice defect* characterized by a closed *Burgers circuit*, in contrast to a *dislocation*. The motion along the circuit is accompanied by a lattice rotation around an axis either parallel to the defect line (known as *wedge disclination*) or perpendicular to it (known as twist disclination). The measure of the lattice distortions associated with a disclination is a dimensionless angle, ω (it is termed the Frank vector), necessary for compensating the rotation accumulated in going along the Burgers circuit. Disclinations with $\omega \cong 0.01$ can occur in the course of *plastic deformation* at *true strains* $\gamma > 1$. Their motion is often accompanied by the occurrence of strongly disoriented *grain parts*. Disclination can be thought of as a border line of a *low-angle boundary* terminating inside a grain.

discontinuous coarsening *Coarsening* of lamellae or rods in *colonies* formed in *discontinuous precipitation*. This type of coarsening proceeds much faster than a common *particle coarsening,* because the former is controlled by *diffusion* over the colony *interface*, whereas the latter is governed by the *bulk diffusion*.

discontinuous dissolution Disappearance of *colonies* formed in *discontinuous precipitation*. This is a result of *phase transformation* upon heating.

discontinuous grain growth See *abnormal grain growth*.

discontinuous precipitation *Decomposition* of a *supersaturated solid solution*, α', by nucleation and growth of two-phase *colonies* of the decomposition products, α and γ, α being the same solid solution but with a much smaller supersaturation, if any, and γ being an *intermediate phase*. The colonies look like *eutectoid colonies* and nucleate at the *grain boundaries* of the parent α'-phase. This phenomenon is also termed *cellular precipitation*.

discontinuous recrystallization *Recrystallization* evolving on *annealing* via nucleation and growth of *strain*-free *grains* (*recrystallization nuclei*) in a *plastically deformed matrix*. See *primary recrystallization*.

discontinuous yielding See *yield-point elongation*.

dislocation *Linear lattice defect* characterized by an open *Burgers circuit*; the *Burgers vector* is a measure of the corresponding lattice distortions and determines *dislocation energy*. There are dislocations (D) of various types, i.e., *edge, screw,* or *mixed,* and of different *sense.* D within crystals form closed *loops* and a *dislocation network.* D occur in the course of *crystallization,* and the *dislocation density* strongly depends upon the *impurity concentration* and *linear growth rate,* on thermal and *concentration* gradients in the growing crystals (see, e.g., *cellullar substructure*), and on the deviation from *stoichiometry* (in *intermediate phases* of a constant composition). D can increase in number in the course of *plastic deformation* (see *dislocation multiplication*), whereas *recovery* and *primary recrystallization* decrease the *dislocation density.* A conservative motion of D (i.e., the motion without any *long-range* atom transport) is referred to as *glide* and results in plastic deformation by *slip* or *deformation twinning.* Long-range *strain* fields are associated with D (see *dislocation stress field*), which is revealed in the interaction of D with *point defects* (see *Cottrell atmosphere* and *climb*) and with other D. For instance, repulsive forces act between parallel D of the same type and sense, lying on the same *slip plane.* Attractive forces occur between D of the same type, but of different senses, lying on the same slip plane; the latter can result in *dislocation annihilation.* If the parallel edge D of the same sense lie on parallel slip planes, they are attracted to each other and form a D wall (see *low-angle boundary*). D of different types can interact in cases in which their stress fields have common components. D can also interact with *planar defects* (see *strain hardening* and *grain-boundary strengthening*). In *ionic crystals,* D distorts the charge balance. In semiconductor *crystals,* edge and mixed D provide *deep centers.*

dislocation annihilation Complete disappearance of attracting *dislocations* of the same type but of the opposite *sense,* provided they are parallel and lie on the same *slip plane.*

dislocation core Volume along a *dislocation* line wherein atomic displacements cannot be described by linear elastic theory. The core radius, $rc \cong 2b$ in *metallic crystals,* is smaller than b in *covalent* and *ionic crystals,* where b is the *Burgers vector.* The *atomic structure* of the dislocation core is not known yet, but as follows from computer simulations, the volume of the dislocation core is close to that of a row of *self-interstitials.*

dislocation creep Creep deformation due to the dislocation *glide* motion and dislocation rearrangement (i.e., *dynamic recovery*) controlled by *pipe diffusion* (at temperatures $0.3–0.5\ T_m$) or by *lattice diffusion* (at temperatures $>0.5\ T_m$). In many *metallic* and *ionic polycrystals,* a *subgrain structure* forms at this creep regime, the subgrain size decreasing with creep *strain.*

dislocation delocalization Spreading of the *core* of a *primary dislocation* trapped by a *general grain boundary.* Delocalization leads to an energy reduction of the *system* "boundary-trapped dislocation" because the core spreading of the latter can be considered as *splitting* into many *grain-boundary dislocations,* whose *Burgers vectors* are so small that the energy of the delocalization products is lower than that of the trapped dislocation. Delocalization distorts the *atomic structure* of the boundary and increases its *energy, diffusivity,* and *mobility.* These changes can sustain for relatively

long periods, especially if the boundary migrates at a significant rate, as,
e.g., in the course of *primary recrystallization* or *abnormal grain growth*.

dislocation density Total length of *dislocation* lines per unit volume. In commercial metallic *alloys*, the dislocation density, ρ_d, varies from ~10^6,
after *recrystallization*, to 10^{11}–10^{12} cm^{-2}, after heavy *plastic deformation*.
In some semiconductor *single crystals* ρ_d can be lower than 10 cm^{-2}. If
dislocations are distributed randomly, the mean distance between them
is ~$\rho_d^{-1/2}$.

dislocation dipole Pair of parallel *edge dislocations* of the same type, but of the
opposite sense (see *dislocation sense*), lying close to one another on
parallel *slip planes*. Dipoles can form in the course of the *glide* motion
of *screw dislocations* with edge-type *jogs*.

dislocation dissociation See *dislocation splitting*.

dislocation energy *Elastic strain energy* of *dislocation* per its unit length,
~$Gb^2/2$, associated with a stress field around the dislocation line, apart
from the unknown energy of *dislocation core* (*G* is the *shear modulus,*
and *b* is the *Burgers vector*). Since the total energy of the dislocation line
reduces with its length decreasing, dislocation energy is also known as
dislocation line tension.

dislocation line tension See *dislocation energy*.

dislocation loop Dislocation line closing itself inside a *grain*, different parts of
the loop being of different character (*edge, mixed,* or *screw*) and of
different *sense,* but of the same *Burgers vector*. If the Burgers vector of
such a loop lies on the loop plane, the loop is *glissile*; in the opposite
case, it is *sessile* (see *prismatic loop*).

dislocation multiplication Increase in *dislocation density*, mostly because of
multiple cross-slip or the action of *dislocation sources*.

dislocation network Three-dimensional network formed by dislocation lines
within a strain-free grain. The dislocations arrange in such a way that they
meet in nodes, where the sum of their *Burgers vectors* equals zero if the
dislocations are directed to the node, and thus form a three-dimensional
network. There can also be a two-dimensional dislocation network as,
e.g., a *low-angle boundary*.

dislocation pinning Hindering of the dislocation *glide* motion by *solute* atoms
(see *solid solution strengthening*) or by disperse particles (see *precipitation strengthening*).

dislocation sense Conventional characteristic of dislocations assigned arbitrarily
to one of them and then fixed for all the others in the *crystal* considered.
For instance, let in a closed *dislocation loop* ABCD of a rectangular shape,
one pair of its parallel sides, AB and CD, be *edge dislocations,* and the
other, BC and DA, *screw* ones (it should be kept in mind that the *Burgers
vector* for all the sides of the loop is the same). If the edge dislocation AB
is assumed positive, then the screw dislocation BC is right-handed, the edge
dislocation CD is negative, and the screw dislocation DA is left-handed.

dislocation source Structural feature emanating *dislocations* under the influence
of an applied force (i.e., during *plastic deformation*) or other stress

sources, as, e.g., *thermal* or *transformation stresses*. Free surfaces, *grain boundaries*, *interfaces*, or specific dislocation configurations inside the grains (see *Frank–Read source*) can act as dislocation sources.

dislocation splitting Dissociation of a *perfect dislocation* into *partial dislocations* and a *stacking fault* ribbon between them. The total energy of the dissociation products is lower than the energy of the original dislocation. The width of the stacking fault ribbon depends on the *stacking-fault energy*. If it is high, the width of the ribbon does not exceed the *dislocation width*. If it is low, the width of the ribbon can be greater than $10b$ (b is the *Burgers vector* of perfect dislocation). Dislocation splitting is also called dislocation dissociation. See also *extended dislocation*.

dislocation stress field *Long-range* field whose energy per unit-length of dislocation is proportional to $Gb^2 ln(R/r_c)$. Here, G is the *shear modulus*, b is the *Burgers vector*, R is the distance from the dislocation line, and r_c is the radius of the *dislocation core*.

dislocation structure See *substructure*.

dislocation tangle Braid of *dislocations* forming at the stage of *multiple slip* and developing into a dislocation cell wall on further deformation (see *cell structure*). The *dislocation density* in tangles reaches $\sim 10^{12}$ cm^{-2}.

dislocation wall See *low-angle boundary*.

dislocation width Width of the area along a dislocation line on the *slip plane*, within which atomic displacements from their *lattice* positions exceed some predetermined limit. Dislocation width decreases with an increase in the *bond energy* and bond anisotropy; in *ionic* and *covalent crystals* it is much lower than in *metallic crystals*.

disordered solid solution *Solid solution* in which *solute* atoms occupy the sites of the *host* lattice randomly (compare with *ordered solid solution*). It is also referred to as random solid solution.

disordering Transformation of an *ordered solid solution* into disordered one. See *order–disorder transformation*.

disorientation Mutual arrangement of two identical *crystal lattices* defined as follows. Take two lattices with one site in common, pass a straight line through the site, and rotate one lattice around the line until all its sites coincide perfectly with all the sites of the other lattice. This line is called the disorientation axis, and the rotation angle is referred to as the disorientation angle. Disorientation is usually denoted by $X°\langle uvw \rangle$, where X is a disorientation angle and $\langle uvw \rangle$ are crystallographic indices of the corresponding axis. Owing to lattice symmetry, the same *disorientation* can be obtained using various pairs' disorientation angle–disorientation axis (in *cubic systems*, there are 24 such pairs). The minimum of the possible angles describing the same diorientation is often taken as a disorientation angle. Thus, the disorientation angle is often assumed to be the minimum of all the possible disorientation angles. Disorientation is also called *misorientation*.

dispersed phase Single-phase *microconstituent* formed by *crystallites* that occur because of *precipitation* and are much smaller than the *matrix grains*.

dispersion strengthening Strength increase caused by *incoherent particles*, in comparison to the *matrix phase*. See *Orowan mechanism*.

dispersoid See *dispersed phase*. This term frequently relates to inert, e.g., oxide, particles in *dispersion-strengthened alloys* or to *incoherent precipitates*.

dispersoid-free zone Narrow, particle-free zone delineating the *grain boundaries* arranged perpendicular to the principal *tensile stress* in the sample. Such zones occur because of *diffusional creep*.

displacement cascade Cluster of various *point defects*, mostly *vacancies* and *self-interstitials*, occurring due to *irradiation damage* by high-energy (~1 MeV) ions or thermal neutrons. The size of the clusters depends on both the mass and energy of the primary particles and on the *atomic mass* in the material irradiated. Cascades are characterized by a high *concentration* of self-interstitials at the cascade periphery and by an identical concentration of vacancies in the cascade core.

displacement shift complete (DSC) lattice Auxiliary lattice used for description of the *atomic structure* of *high-angle grain boundaries*. Planes of DSC lattice are parallel to the *CSL* planes and pass through all the atomic sites in the *crystal lattices* of two disoriented *grains*. The shift of the CSL along any *translation vector* of the DSC lattice changes only the CSL position, and does not change the CSL itself. Thus, these vectors can be considered to be the *Burgers vectors* of perfect *grain-boundary dislocations*. The minimum length of the vector equals b/Σ, where b is the Burgers vector of *primary dislocation*, and Σ is a CSL parameter.

displacive transformation Name of *shear-type transformation* used in ceramic science.

divacancy Complex of two *vacancies* in adjacent *lattice sites*. Divacancy has a smaller *free energy* in comparison to two single vacancies far removed from one another.

divorced eutectoid *Microconstituent* that, in *binary systems*, is supposed to be two-*phase* (see *eutectoid colony*), but, in fact, is single-phase. Such a structure forms when *crystallites* of the occurring solid phases grow independently, and crystallites of one of these phases join up with *proeutectoid* crystals of the same phase. Divorced eutectoid is also observed when the volume fraction of a phase undergoing *eutectoid reaction* is low (<5 vol%).

divorced pearlite Divorced eutectoid in steels.

dodecahedral plane {110} plane in *cubic structures*.

domain structure In *ferromagnetic* and *ferrimagnetic* materials, a *structure* formed by *magnetic domains;* in *ferroelectrics,* it is formed by *ferroelectric domains*. Domain structure decreases the energy of magnetic (or electric) poles at the free surface of a body, at the *grain boundaries* and *interfaces*, at particles whose size is greater than the *domain wall* thickness, etc. The energy is known as magnetostatic or electrostatic energy, respectively. In the absence of an external magnetic (electric) field, domain structure eliminates the poles completely, and the magnetization (polarization) vectors of various domains are oriented in such a way that their sum over the body's volume is zero. In *single-domain particles*, magnetic domain structure is lacking. An increase of an external magnetic (or electric) field is accompanied by changes in domain structure proceeded by *domain wall* displacements.

domain wall Layer between two *magnetic* (or *ferroelectric*) *domains* wherein the magnetization (polarization) vector rotates from the position in one domain to that in the neighboring one. If the rotation angle equals 180°, the domain wall is called 180 wall; if 90°, then 90 wall, etc. The wall thickness of magnetic domains depends on its type (see *Bloch wall* and *Néel wall*), on the constant of *magnetic crystalline anisotropy,* and on the energy of *exchange interaction*. In ferroelectrics, the thickness of domain wall is ~1 nm. *Grain* and *phase boundaries*, *second-phase* particles, nonhomogeneous *microstrains*, pores, etc. inhibit displacement of domain walls, with an accompanying increase in external magnetic (electric) field. This, in turn, affects the shape and area of the hysteresis loop, as well as other magnetic (ferroelectric) properties.

donor *Dopant* in semiconductors increasing the concentration of charge carriers. The energy of the donor valence electrons lies inside the *band gap* close to its top. Owing to this, the valence electrons can be activated to reach the *conduction band* and take part in conductivity. For instance in elemental semiconductors (Si, Ge), donors can be *substitutional solutes* with higher valence than *host* atoms as well as *interstitial solutes*.

dopant In *crystalline ceramics*, a dopant traditionally means the same as an *alloying element* in *metallic alloys*. In semiconductors, dopant is a *component* intentionally introduced in strongly controlled amounts for increasing the concentration of charge carriers (see *donor* and *acceptor*).

doping Intentional addition of some *components*, known as dopants, with the aim of affecting properties of the material doped. In semiconductors, dopants are introduced in restricted and strongly controlled amounts. In *crystalline ceramics*, dopants play the same role as do *alloying elements* in metallic *alloys*. Microalloying of metallic materials, i.e., alloying with small amounts of alloying elements, is also termed doping. In particulate *composites*, doping means an addition of chemically neutral *dispersoids* as, e.g., Y_2O_3 in *ODS* nickel alloys.

double aging Two-stage *precipitation treatment* increasing homogeneity in the arrangement of *precipitates*. At the first stage, an almost-homogeneous distribution of *GP zones* is produced, and at the second stage, at a temperature above the *solvus* for GP zones, they transform into *second-phase* precipitates and are distributed more homogeneously than after a one-stage *aging treatment*. See *rule of stages*.

double cross-slip *Glide* motion of a part of a *screw dislocation* from one *slip plane* to another crossing the first one (see *cross-slip*) and then to a third one parallel to the first slip plane. The *zone axis* of the intersecting planes must be parallel to the *Burgers vector* of the dislocation. Double cross-slip is one of the mechanisms of *dislocation multiplication* during *plastic deformation*.

double kink See *kink*.

double stacking fault See *extrinsic stacking fault*.

doublet Two components of a set of *characteristic x-rays* with a small difference in wavelengths (as, e.g., $K\alpha_1$ and $K\alpha_2$ in the K set).

drag force Thermodynamic force slowing down *grain boundary* migration by decreasing the *driving force*. Drag force is always a reaction force acting on a moving boundary and cannot induce migration of a stationary boundary. Units of drag force are the same as those of driving force. See *groove drag*, *impurity drag,* and *particle drag*.

drift In materials science, directed atomic flux superimposed on a basic, oriented or random flux. See, e.g., *uphill diffusion*.

driving force Difference in the *free energy* of a *system* between its initial and final configurations. Driving forces promote *phase transitions* as well as *microstructural* changes not connected with such transitions as, e.g., *primary recrystallization, Ostwald ripening, grain growth*, etc. The units of driving force are J/m,^{3} equivalent to N/m.^{2}

dual-phase microstructure Structure consisting of two single-phase *microconstituents* with comparable *grain sizes*. In dual-phase microstructures, one of the microconstituents is of a lower volume fraction and its grains are arranged along the *grain boundaries* of the other.

ductile-brittle transition [temperature] Temperature at which a numerical value of some ductility characteristic decreases by 50% of its original value upon the temperature lowering. A decrease in the temperature corresponds to an increase in the material toughness.

ductile cast iron See *nodular cast iron*.

ductility transition [temperature] See *ductile-brittle transition*.

duplex grain size Inhomogeneity of a single-phase *microstructure* revealing itself in the presence of two grain populations characterized by strongly different *modes*.

duplex microstructure *Structure* consisting of two single-phase *microconstituents* of nearly equal volume fractions; their *grains* are arranged randomly and have nearly the same *mean size*.

dynamic recovery Decrease of *dislocation density* in the course of *plastic deformation*. This is a result of dislocation *climb* and *multiple cross-slip* that can lead to the formation and growth of *subgrains*. The rate of dynamic recovery depends on the deformation rate and the temperature and is greater than that of *static recovery*. Softening due to dynamic recovery, on the one hand, and *strain hardening* due to deformation, on the other, can compensate each other to such a degree that *superplasticity* is revealed.

dynamic recrystallization *Primary recrystallization* evolving concurrently with the *strain hardening* during *hot deformation*. A certain *true strain* (referred to as *critical deformation*) is necessary for dynamic recrystallization (DR) to start. The applied load deforms *recrystallized* grains occurring in the course of DR and decreases their growth ability. Simultaneously, new recrystallized grains nucleate and grow in other areas of the deformed matrix until their *driving force* reduces as a result of their deformation. Thus, unlike *static recrystallization*, DR does not evolve steadily, but commences sporadically, in different areas of the deformed specimen, and then stops. When softening due to DR counterbalances *strain hardening*, the further *plastic deformation* is not accompanied by strengthening. *The mean grain size, $\overline{D}$, at this stage is:*

$$\overline{D} = k \cdot \sigma_s^n$$

where k is a constant, σ_s is the *flow stress* at the stage, and $0.5 \leq n \leq 1$. $\overline{D}$ is smaller than the mean grain size after comparable *cold deformation* and static recrystallization. If cooling after hot deformation is rapid, some *stored energy* is retained in the deformed material, and *annealing* upon cooling results in either *metadynamic* or *postdynamic recrystallization*.

dynamic strain aging Sporadic inhibition of *plastic deformation* at increased temperatures and low strain rates, owing to a dynamic equilibrium between *dislocation pinning* by *Cottrell atmospheres* and their unpinning under the influence of *thermal activation*. This manifests itself in a serrated *stress–deformation curve*. Dynamic strain aging is analogous to *strain aging* in the sense that both result from an interaction of dislocations with *interstitial solutes*, but the interaction in the former takes place in the course of deformation, whereas in the latter case, it evolves after the deformation. Dynamic strain aging is also referred to as the *Portevin–Le Chatelier effect*.

E

ε-carbide *Transient phase* of an approximate *composition* $Fe_{2.4}C$ with the *hexagonal crystal structure* precipitating in *steels* from as-quenched *martensite* at a low-temperature *tempering treatment*.

ε-martensite Martensite with *HCP crystal structure* occurring either at high external pressures or in some *plastically deformed austenitic steels*. Its occurrence may be connected with the existence of *HCP allotropic modification* of iron, known as ε-Fe.

η-carbide *Transient phase* of *composition* Fe_2C with *orthorhombic crystal structure*; it precipitates in *steels* from as-quenched *martensite* at a low-temperature *tempering treatment* (at 50–200°C).

earing Phenomenon revealing itself in the appearance of undulations along the rim of deep drawn caps and connected with *planar anisotropy* of the sheet used and, thus, with its *crystallographic texture*. The propensity for earing can be estimated from the magnitude of *Δr-value*.

easy glide Stage of *plastic deformation* in *single crystals* wherein only one *slip system* is active and the *strain hardening* rate is low due to a small increase in *dislocation density*. In crystals with *HCP structure*, such a stage can be observed up to relatively high *strains*. In crystals with *cubic crystal structure*, this stage is either rapidly relieved by *multiple slip* or is not observed at all.

easy magnetization direction *Lattice direction* along which the magnetization energy in *ferromagnetic* or *ferrimagnetic single crystals* is minimum. In Fe, it is ⟨001⟩; in Ni, it is ⟨111⟩; and in Co, it is ⟨0001⟩. Magnetization vectors in *magnetic domains* lie close to or coincide with the easy magnetization direction.

edge dislocation Dislocation whose *Burgers vector* is perpendicular to its line. It can be considered the edge of an extra plane inserted in a *crystal lattice*. If the extra plane is situated above the *slip plane* of the dislocation, the dislocation is assumed positive; in the opposite case, it is assumed negative. It should be noted that the *dislocation sense* is a conventional characteristic.

elastic deformation Deformation induced by some external force and disappearing after its removal. Elastic deformation can also be associated with *residual stresses* or with crystal defects and their agglomerations (e.g., *coherency strain*), etc.

elastic modulus Coefficient of proportionality between the *stress* and *strain* in the elastic regime of loading (see *Hooke's law*). Its units are MN/m^2 (MPa) or GN/m^2 (GPa), i.e., the same as those of *stress*. Elastic modulus in *single crystals* is *anisotropic*, and its magnitude for *polycrystals* given in literature relates to materials without *crystallographic texture*. See *Young's modulus*, *shear modulus*, and *bulk modulus*.

elastic scattering Scattering of x-ray (or electron) radiation by electrons in solids, the scattered radiation being of the same wavelength (energy) as the primary radiation. The term "coherent scattering," meaning the same thing, was used in the past.

elastic strain energy Amount of *free energy*, per unit volume, associated with *elastic deformation*. In the case of *tension*, it equals $E\varepsilon^2/2$, where E is *Young's modulus* and ε is the *strain*.

electro-etching Electrochemical etching of *metallographic samples* used for some corrosion-resistant materials; also used in cases in which *chemical etching* produces a rough surface topography.

electromigration Mass transport under the influence of a gradient of electrical potential. It is observed in the conducting lines of microelectronic devices where the current density is exceptionally high. Electromigration manifests itself in the formation of voids and hillocks leading to short circuits.

electron:atom ratio See *electron concentration*.

electron backscattered pattern (EBSP) See *electron channeling*.

electron channeling Channeling is a motion of charged particles parallel to some atomic rows in a *crystal*. In the course of this motion, the particles penetrate deep into the crystal. Electron channeling reveals itself in the dependence of the yield of *backscattered electrons* on the *glancing angle* of the primary beam. If this angle equals the *Bragg angle* (for ~20 keV electrons, it is 1–2°), the yield of backscattered electrons strongly decreases. As a result, backscattered electrons produce the image of a crystal *lattice* known as channeling pattern. It is used for determining the orientation of *single crystals* by electron channeling pattern (ECP), or of relatively small (several μm in diameter) *grains* by selected area channeling patterns (SACP).

electron channeling pattern (ECP) See *electron channeling*.

electron compound/phase See *Hume–Rothery phase*.

electron concentration The number of valence electrons per atom affecting *crystal structure* of certain *intermediate phases* (see, e.g., *Hume–Rothery phases*). Electron concentration is also known as electron:atom ratio.

electron diffraction *Elastic scattering* of electrons, identical to *x-ray diffraction*, and obeying *Bragg's law*. Due to a high *atomic scattering factor* for electrons in comparison to x-rays, *substructure* images (see *diffraction contrast*) and *electron diffraction patterns* can be obtained from a small amount of substance, e.g., from *thin foils* or small particles.

electron diffraction pattern Pattern (consisting mostly of *diffraction spots*) used for determining the *crystal structure* or orientation of *(sub)grains* and *precipitates*, or the periodic *atomic* structure of free surfaces (see, e.g., *LEED*).

electronegativity Quantity characterizing the ability of an atom to attract electrons. Its magnitude is close to the sum of the energy necessary to add an electron to a neutral atom and the energy necessary to remove an electron from a neutral atom. If the electronegativity difference of various atoms in a compound is large, their *bond* is, for the most part, *ionic*; in the opposite case, it is largely *covalent*. See also *Hume–Rothery rules* for solubility in *substitutional solid solutions*.

electron energy loss spectroscopy (EELS) Technique for determining a material's chemistry. It utilizes the changes in the energy of transmitting electrons that interact with the inner electron shells of atoms in the specimen. EELS can be used for qualitative and semiquantitative chemical analyses.

electron micrograph Photograph of *microstructure* obtained by *PEEM*, *SEM*, or *TEM*.

electron [micro]probe Focused electron beam used for studying the *crystal structure* as well as for chemical analysis.

electron microscopy (EM) Technique for studying *substructure, crystal lattice*, and orientation of small particles or *subgrains*, different *crystal defects* (*TEM, HVEM*, and *HRTEM*), as well as topography of the specimen surface (*SEM*).

electron probe microanalysis (EPMA) Technique for elemental chemical analysis wherein the primary electron beam excites the *characteristic x-rays* in the specimen. The energy (or wavelength) of these x-rays is used as a sensitive indicator of the *composition* in the area studied. The *resolution* limit of the technique is 1–5 μm on the surface, as well as into the depth. Scanning of the specimen surface provides data about distribution of various chemical elements.

electron spectrometry for chemical analysis (ESCA) Technique for qualitative and quantitative chemical analyses of thin (about 5–10 *interatomic spacings*) surface layers by irradiating the specimen with *monochromatic* x-rays and measuring the kinetic energy of ejected photoelectrons.

elementary jog See *jog*.

embryo In materials science, a small area inside a parent phase characterized by a different *atomic structure* and, frequently, by a different chemical *composition*. Embryos occur due to thermal *fluctuations* aided by *strain* fields and *segregations* in the solid parent phase. The embryo size is smaller than that of the *critical nucleus*, thus, the embryos are thermodynamically unstable.

energy-dispersive diffractometry (EDS/EDAX) Technique for *x-ray structure analysis* using a *diffractometer* in which a *polycrystalline* specimen and an energy-dispersive counter are fixed at specified angles, i.e., the specimen at some angle Ψ and the counter at the angle 2Ψ, with respect to the primary beam of *white* radiation. In the beam, there are always x-rays whose energy corresponds to the *Brag equation* for at least the majority of the *phases* in the specimen. An analysis of the *crystal structure* of the phases can be done in a much shorter time than in the case of the conventional *diffractometric method.*

energy-dispersive spectrometry Technique for chemical analysis by spectrometry of *x-ray fluorescent* radiation produced under the influence of a primary high-energy electron beam. The x-rays emitted by a fixed sample are directed to a fixed energy-dispersive counter. The counter is connected to a multichannel analyzer that classifies the emitted x-rays and yields their *energy spectrum.* This spectrum can be used for the qualitative and quantitative chemical analyses.

energy spectrum Distribution of the x-ray intensity or of the number of electrons (ions) versus their energy.

engineering strain See *nominal strain.*

engineering stress See *nominal stress.*

epitaxial dislocation *Misfit dislocation* at the *interface* between an *epitaxial film* and its substrate.

epitaxial film Single-*crystalline* film characterized by some orientation relative its substrate (see *epitaxy*). The *orientation relationship* between the film and substrate is described primarily by parallelism of *close-packed directions* in their *lattices.* There can be *homoepitaxial* and *heteroepitaxial* films.

epitaxy Oriented growth of a deposit on a *crystalline* substrate accompanied by the occurrence of an *orientation relationship* between the deposit and the substrate. It is called homoepitaxy if the deposit and the substrate are of the same chemical *composition* and identical *crystal structure.* In the opposite case, it is called heteroepitaxy. See also *epitaxial film, homoepitaxial film,* and *heteroepitaxial film.*

equatorial net See *polar net.*

equiaxed Having nearly the same size in all directions.

equilibrium diagram See *phase diagram.*

equilibrium phase Phase whose range of existence corresponds to the minimum *free energy* of the *system* concerned. It is also referred to as stable phase.

equilibrium segregation Increased *concentration* of *solute* atoms at *crystal defects,* such as free surface, *grain boundary, interface, stacking fault,* or *dislocation,* in comparison to the surrounding *solid solution,* which lowers the *free energy* of the whole *system.* See, e.g., *Cottrell* and *Suzuki atmospheres* and *grain-boundary segregation.*

equilibrium system Body, or an aggregate of bodies that does not exchange the energy and substance with the environment (it is called an isolated system) and has the minimum *free energy.* See also *thermodynamic equilibrium* and *system.*

equilibrium temperature In materials science, the temperature, T_0, at which the parent and new *phases* are at equilibrium. As seen in the *phase*

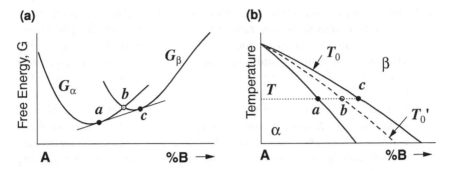

FIGURE E.1 (a) Free energies, G_α and G_β, of α and β solid solutions versus concentration at temperature T. The tangent corresponds to the constancy of thermodynamic potentials of components in α- and β-phases, and points a and c show their compositions at equilibrium. Intersection b corresponds to the equilibrium of metastable solutions of the identical composition. (b) Part of a binary phase diagram and corresponding equilibrium temperatures, T_0 and T_0'. See text.

diagram (see Figure E.1), the equilibrium *compositions* of the α- and β-phases, a and c, at any temperature T_i can be found with the aid of a tangent to the curves $G_\alpha(x)$ and $G_\beta(x)$, where $G_\alpha(x)$ and $G_\beta(x)$ are the composition dependences of the *free energy* of the α- and β-phase, respectively. At the same temperature T_i, there can be the equilibrium of *metastable* α- and β-phases of an identical composition, b. A locus of points, b_i, corresponding to different temperatures T_i displays the equilibrium temperature, T_0', of metastable phases. A typical example of such equilibrium is observed in *martensitic transformation*; in this case, the *undercooling* necessary for the *martensite* nucleation equals $T_0' - M_s$.

equivalence diagram Presentation of the phase equilibrium in *systems* with *ionic bonds* in an equivalent concentration space. It is analogous to a *phase diagram*, but the *components* are chosen in such a way that varying their proportion does not disturb the charge neutrality. For instance, in the equivalence diagram for quaternary *sialons*, the components are Si_3O_6, Al_4O_6, Al_4N_4, and Si_3N_4, instead of SiO_2, Al_2O_3, AlN, and Si_3N_4, respectively, as in the common phase diagram. This leads to the following formulae for equivalent%:

$$\text{eq.\% Al} = 3[\text{Al}]/(4[\text{Si}] + 3[\text{Al}]); \quad \text{eq.\% O} = 2[\text{O}]/(3[\text{N}] + 2[\text{O}])$$

where the symbols in the square brackets are contents of Al, Si, N, and O. The fact that the component *concentrations* are connected in the formulae decreases the number of the *degrees of freedom* by 1, and so the equivalence diagram for the quaternary system is identical to an isothermal section of a *ternary* diagram. Equivalence diagram is also called compatibility diagram.

etch figure Pit on the specimen surface having well-formed flat facets coinciding with *densely packed lattice planes*. It can be obtained by *chemical, thermal,* or *electro-etching*. Etch figures are used for determining orientation of *grains* (minimum diameter ~10 μm) or *single crystals* with the aid of a goniometric *optical microscope* (with an accuracy of 0.5–1°).

etch pit Pit on an electropolished specimen surface occurring upon etching. If certain special etchants are used, pits appear at the intersections of *dislocations* with the specimen surface. In this case, the etch pits can be used for determining the *dislocation density* (at $\rho_d < 10^8$ cm^{-2}) and for studying the dislocation arrangement.

Euler angles Rotation angles φ_1, Φ, and φ_2 necessary for the coordinate system of a specimen to become coincident with the coordinate system of a *crystallite*. If both of the systems are orthonormal, as, e.g., in the case of a rolled sheet sample and a material with *cubic lattice*, the Euler angles can be defined as follows. Let both the systems have a common origin. Rotate the sample coordinate system around *ND* by an angle φ_1 until *RD* reaches the (001) plane of the crystallite. Then, rotate around RD (in its new position) by an angle, Φ, until ND reaches [001]. Finally, rotate around ND (in its last position) by φ_2 until *TD* reaches [010]. For instance, $\varphi_1 = 0°$, $\Phi = 0°$, and $\varphi_2 = 0°$ for (100) [001] orientation, and 0°, 45°, and 0°, respectively, for (110) [001]. The Euler angles form a three-dimensional space (known as Euler space) used for displaying the *orientation* and *misorientation distribution functions*.

eutectic colony Smilar to an *eutectoid colony*, but appearing in *eutectic reactions*.

eutectic point Chemical *composition* of a liquid *phase* taking part in a *eutectic reaction* (e.g., point x in *binary* system [see Figure E.2]).

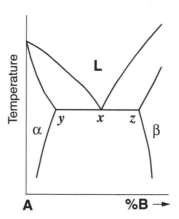

FIGURE E.2 Part of a binary phase diagram with an eutectic reaction. In the case of an eutectoid reaction, the phase fields are of the same configuration, but instead of a liquid phase, *L*, there should be some solid phase, *γ*.

eutectic reaction *Phase transformation* in a *binary* system (see Figure E.2) following the reaction:

$$L_x \leftrightarrow \alpha_y + \beta_z$$

where L_x is a liquid phase of *composition* x, and α_y and β_z are solid phases of compositions y and z, respectively; the right end of the arrow shows the reaction path on cooling, and the left one shows the path on heating. In *ternary* systems, eutectic reactions can evolve as follows:

$$L_a \leftrightarrow \alpha_b + \beta_c + \delta_f$$

where L is a liquid and α, β, and δ are solid phases, and the indices relate to the compositions of the participating phases, or as:

$$L \leftrightarrow \gamma + \eta$$

where L is a liquid and γ and η are solid phases. According to the *Gibbs phase rule*, the first two reactions are *invariant,* thus, the temperature, pressure, and phase compositions remain constant. In the third one, phase compositions vary in the course of the reaction.

eutectic [structure] *Microconstituent* appearing due to an *eutectic reaction* and usually evolving by the nucleation and growth of *eutectic colonies.* In some *alloys*, e.g., in *gray cast-irons*, this typical eutectic structure cannot be detected by *optical microscopy* (eutectics in these cases are termed irregular). The fraction of the eutectic constituent equals the fraction of the liquid phase whose *composition* corresponds to the *eutectic point* (see *lever rule*).

eutectic temperature Temperature of a *eutectic reaction* in a *phase diagram.*

eutectoid colony Multiphase complex consisting of interpenetrating *crystallites* of the *phases* occurring in *eutectoid decomposition* and growing together into the parent phase. The nucleation and growth of one of the crystallites helps the others to nucleate and grow (see *coupled growth*), which eventually leads to the formation of a colony and to an *orientation relationship* between the crystallites in the colony.

eutectoid decomposition *Eutectoid reaction* upon cooling.

eutectoid point Chemical *composition* of a solid phase decomposing in *eutectoid reaction* upon cooling (or that occurring upon heating).

eutectoid reaction *Phase transition* following the reaction:

$$\gamma_x \leftrightarrow \alpha_y + \beta_z$$

where γ_x, α_y, and β_z, are solid phases of *compositions* x, y, and z, respectively (see Figure E.2); the right end of the arrow shows the reaction path upon cooling (known as eutectoid decomposition), and the left one shows the reaction path upon heating. Independently of the reaction path, this reaction in *binary systems*, according to the *Gibbs phase rule*, is *invariant* and evolves at a constant temperature and pressure and constant x, y, and z.

eutectoid [structure] *Microconstituent* occurring on *eutectoid decomposition* (e.g., *pearlite* in *steels*) and formed by *eutectoid colonies*. The fraction of the eutectoid constituent equals the fraction of the decomposing parent phase (see *lever rule*). In some cases, especially when the eutectoid volume fraction is low, no colonies can form; in this case, this microconstituent is called *divorced eutectoid*.

eutectoid temperature Temperature of an *eutectoid reaction* in a *phase diagram*.

Ewald sphere Sphere of the radius $1/\lambda$, where λ is the wavelength. This geometric construction is used to solve various problems of *electron (x-ray) diffraction* with the aid of *reciprocal lattices*. If the origin 0 0 0 of a reciprocal lattice lies on the sphere, then the site of the reciprocal lattice corresponding to the reflecting plane of the *crystal lattice* should also lie on the sphere. Ewald sphere is also referred to as reflection sphere.

exaggerated grain growth See *abnormal grain growth*.

excess free volume Difference in the *specific volumes* of an *amorphous phase* and a *crystalline phase* of the same *composition*. The same term is also applied to *crystal defects* because the *atomic volume*, e.g., of *grain* and *phase boundaries*, is greater than that of a perfect crystal.

exchange interaction Quantum-mechanical interaction aligning spin magnetic moments of the neighboring atoms of transition elements either parallel (in this case, the energy of the interaction is assumed positive) or antiparallel to one another (negative exchange interaction). In the first case, the substance is *ferromagnetic*, and in the second one, *ferrimagnetic* or *antiferromagnetic*. In some ferromagnetics and ferrimagnetics containing rare-earth elements, the alignment of elementary magnetic moments is not strongly parallel or antiparallel. Interaction between neighboring atoms is named direct exchange interaction. In some cases, e.g., in *ortho-ferrites*, magnetic coupling results from superexchange interaction involving, along with metallic ions, also O^{2-} ions.

extended dislocation *Splitted dislocation* wherein *partial dislocations* are separated by a narrow *stacking fault* ribbon.

extinction Decrease in the *intensity* of a diffracted x-ray (electron) beam due to interaction of the primary x-rays (electrons) with the scattered ones inside a sample. X-ray extinction is observed at an increased *subgrain size* and reaches its maximum in an ideally perfect *single crystal* (see *primary* and *secondary extinction*). Electron extinction can be observed when a foil is bent (see *extinction contour*) or when there is an *interface* inclined to the foil surface (see *thickness fringes*).

extinction coefficient See *absorption coefficient*.

extinction contour In *bright-field TEM* images, a dark band occurring due to the foil bending, which changes the orientation of *crystal planes* and brings them in the reflecting position. In contrast to *thickness fringes*, extinction contours are irregular in shape and can be arranged arbitrarily. Extinction contours are also termed bend contours.

extinction rule Combinations of the indices of *lattice planes* whose *structure factor* equals zero. For instance, in *BCC structure*, the sum of the plane indices $(h + k + l)$ should be odd, whereas in *FCC structure*, the plane indices are mixed, i.e., both even and odd. Because of this, the *Miller indices* of reflecting planes with the smallest *Bragg angle* in BCC structure are {200}, whereas in FCC, they are {100}.

extraction replica *Direct replica* that retains small *precipitates* extracted from the sample surface layer. To strip off the replica, the sample should be dissolved.

extrinsic grain-boundary dislocation See *grain-boundary dislocation.*

extrinsic stacking fault Disturbance in the stacking sequence of the *close-packed planes* caused by the insertion of an additional plane. For instance, in an *FCC structure* where a perfect stacking sequence is ...ABCAB-CABC... the insertion of an additional plane *A* results in the sequence ...ABCA*BAC*ABC... wherein the layers' arrangement *BAC* is erroneous. The energy of an extrinsic stacking fault is a little higher than that of an *intrinsic* one. Extrinsic stacking fault is also referred to as double stacking fault.

F

face-centered cubic (FCC) structure One of the two most densely packed crystal structures (see Figure F.1) with *coordination number* 12, *atomic packing factor* 0.74, and the *close-packed planes* {111} and *close-packed directions* ⟨110⟩. The radius of *tetrahedral void* equals 0.225R and that of *octahedral void* is 0.415R, where R is the *atomic radius*. In a densely packed structure, the atoms of each most densely packed layer lie in the hollows of the lower layer. In the FCC structure, the atom centers of the second layer are arranged in the B hollows of the first layer (see Figure F.2). The atom centers of the third layer are arranged in the hollows of the second layer and above the C hollows of the first layer. The atom centers of the fourth layer are arranged in the hollows of the third layer and above the A centers of the first layer. Thus, the stacking sequence of the atom layers {111} in FCC structure can be described as ...*ABCABCA*....

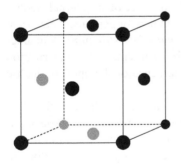

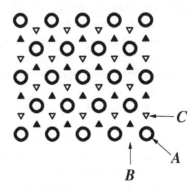

FIGURE F.1 Unit cell of FCC lattice.

FIGURE F.2 Sequence of close-packed planes {111} in FCC lattice. See text.

face-centered lattice *Cubic* or *orthorhombic Bravais lattice* wherein, along with the points at the vertices of the corresponding *unit cell*, there are additional points at the centers of all the cell faces.

F-center *Color center* occurring due to electron trapping by a *vacancy* in a cation *sublattice*.

Fe–C system *Alloy system* with *stable* solid *phases: ferrite, austenite, δ-ferrite,* and *graphite,* the latter being the only carbon-rich phase.

Fe–Fe₃C system. Fe–C *alloy system* with *stable δ-ferrite, ferrite, metastable austenite,* and *cementite,* the latter being the only carbon-rich *phase.*

ferrimagnetic. Material revealing *spontaneous* magnetization below *Curie point,* T_C, and characterized by negative energy of *exchange interaction* and by unequal and oppositely directed magnetic moments of different magnetic *sublattices.* The magnetic behavior of ferrimagnetics below T_C is similar to that of *ferromagnetics.* However, the temperature dependence of spontaneous magnetization in ferrimagnetic and ferromagnetic substances can be quite different.

ferrite. *Solid solution* of *alloying elements* and/or carbon in α-Fe. The same name is used for *polycrystalline ferrimagnetic* oxide *ceramics* (see *spinel ferrite, hexagonal ferrite, orthoferrite,* and *garnet ferrite*).

ferrite-stabilizer. *Alloying element* increasing the *thermodynamic stability* of *ferrite.* This is most often accompanied by an increase of the A_3 temperature and a decrease of A_4 in *binary alloys* Fe–M. Since ferrite-stabilizers expand the α-phase field in the corresponding *phase diagram,* ferrite can become stable at certain alloy *compositions,* even at the room temperature, as, e.g., in *ferritic steels.*

ferritic [cast] iron. *Gray iron* whose *microstructure* consists of *flake graphite* and a ferritic *matrix.*

ferritic steel. *Alloy steel* consisting, upon *normalizing,* predominately of *ferrite.*

ferroelectric. Material electrically polarized in the absence of an external electric field, which results from a *spontaneous* alignment of permanent electrical dipoles in its *crystal structure* (see, e.g., *perovskite structure*). A characteristic feature of ferroelectrics is their ability to change the polarization direction to the opposite one under the influence of the applied electric field.

ferroelectric domain. *Grain* part of ferroelectric material in which all the electrical dipoles are parallel to one another at temperatures below *Curie point.* As in *magnetic domains,* polarization directions in the neighboring ferroelectric domains can be antiparallel or perpendicular, and the domain walls are, accordingly, 180° and 90° ones. The thickness of the walls is ~1 nm. See *domain structure.*

ferromagnetic. Material *spontaneously* magnetized below *Curie point,* T_C, due to the parallel alignment of the atomic magnetic moments under the influence of *exchange interaction.* The preferred orientation of the moments decreases with in increase in temperature, and disappears above T_C. In some ferromagnetics containing rare-earth elements, elementary magnetic moments are not strongly parallel.

fiber texture. *Preferred grain orientation* wherein a certain *lattice direction* in the majority of the grains is oriented parallel to a definite direction in the specimen, e.g., parallel to the wire axis or to the normal film surface.

Fick's first law Description of the *diffusion* process at a time-independent *concentration* gradient. A diffusion flux, J_B, of B atoms at a constant temperature is:

$$J_B = -D\Delta c_B / \Delta x$$

where D is a *diffusion coefficient* dependent upon temperature, c_B is the *concentration* of B atoms, and x is the distance. The negative sign in the right-hand part shows that the diffusion flux reduces the concentration gradient.

Fick's second law Description of the *diffusion* process at a time-dependent *concentration* gradient. In cases in which the *diffusion coefficient*, D, is independent of concentration,

$$\partial c_B / \partial t = D(\partial^2 c_B / \partial x^2)$$

where c_B is the concentration of B atoms; in the opposite case,

$$\partial c_B / \partial t = \partial [D(\partial c_B / \partial x)] / \partial x$$

field diaphragm In *optical microscopes*, a diaphragm restricting the *field-of-view* and affecting the image contrast.

field emission Electron emission by an unheated solid in response to a high-voltage electric field. It is also called autoelectronic emission.

field-ion microscope (FIM) Device for the direct observation of *crystal structure*. Atoms of He or Ne are ionized in the immediate vicinity of the sample tip and repelled by the *lattice* ions to a screen yielding an image of the *atomic structure* of the tip. The *magnification* in FIM is $\sim 10^6$.

field-of-view Sample area observed with a certain objective or ocular set. It decreases with an increase in *magnification*, and can also be restricted by a *field diaphragm*.

fine-grained [Material] characterized by a *grain size* of 1–5 µm.

fine pearlite Pearlite with a decreased interlamellar spacing; it occurs at a low-temperature limit of *pearlitic range*. It is also termed sorbite.

fine structure See *substructure*.

firing High-temperature *annealing* of a powder compact resulting in its *sintering*.

first-order transition *Phase transformation* (e.g., *polymorphic transformation*, *eutectoid* or *peritectoid reactions*, *precipitation*, etc.) accompanied by a discontinuous change in *free energy*. In *systems* with first-order transition, both the parent and the new phases coexist in the course of the transition. This means that first-order transitions evolve according to the *nucleation* and growth of the new phase.

first-order twin *Grain* part having a twin *orientation* with respect to the remainder of the grain. The term relates to *annealing* (or growth) *twins* and is used if there are *second-order,* etc. twins.

flake graphite *Graphite crystallites* in *gray cast irons* appearing as relatively thin flakes, curved in different directions, and seemingly divorced from each other. However, in fact, some flakes can grow from a single node and so should be considered strongly disoriented branches of a crab-like crystallite.

flow stress Stress (usually, but not always, *true stress*) necessary to start or to further evolve *plastic deformation*.

fluctuation Incidental and temporal local deviation from an average value. In materials science, a fluctuation is a deviation in *atomic structure* or in *composition*, restricted to a small group of atoms. Fluctuations are associated with *spontaneous* deviations of *free energy* from its average value. The frequency of fluctuations decreases exponentially with an increase in magnitude. Fluctuations increase in magnitude and frequency in the vicinity of *critical points*.

fluorite Name of *ionic* compound of composition CaF_2.

fluorite [structure] type *Crystal structure* identical to the *CaF_2 structure*; it is typical of many *ionic crystals*, e.g., ZrO_2, CeO_2, UO_2, etc.

forbidden gap See *band gap*.

foreign atom Individual atom different from *host* atoms.

forest dislocation Dislocation intersecting an *active slip system*. As a result, dislocations that *glide* over the active system can intersect forest dislocations. Since these intersections are accompanied by the formation of *jogs*, and because the jogs inhibit the glide of *screw* dislocations, forest dislocations contribute to *strain hardening*.

form In crystallography, a family of all the *lattice directions* characterized by the same indices u, v, and w, independent of their signs and the order of their arrangement. It is denoted by placing the indices in angle brackets $\langle uvw \rangle$. A direction of the family is referred to as a direction of the form $\langle uvw \rangle$. If a four-number notation is used, the designation in angle brackets relates to the permutations of the first three indices only; e.g., the form of the *close-packed directions* is denoted by $\langle 11\bar{2}0 \rangle$. A family of all the *lattice planes* characterized by the same *Miller indices* h, k, and l, independent of their signs and the order of their arrangement, is also named a form. Such a family is denoted by indices in braces $\{hkl\}$, and a plane of the family is referred to as a plane of the form $\{hkl\}$. If the *Miller–Bravais* notation is used, the designation in braces relates to the permutations of the first three indices only; e.g., the family of the first-order *prism planes* is denoted by $\{1\bar{1}00\}$.

fragmentation In materials science, an obsolete term relating to the *substructure* formation in the course of *plastic deformation*.

Frank partial dislocation Loop of a *partial dislocation* bordering a *stacking fault* in *FCC structure*. The *Burgers vector* of the Frank partial dislocation is perpendicular to the loop plane, thus, the loop is *sessile*. This dislocation occurs due to the instability of a disc of *vacancies* or *self-interstitials* that collapses at some critical size. See *irradiation damage* and *quench-in vacancies*.

Frank–Read source Element of a *dislocation network* able to produce new
dislocations in the course of *plastic deformation*. The source works as
follows. Consider a segment of a positive *edge dislocation* of length L
lying on its *slip plane* (see stage I in Figure F.3). Let the *resolved shear
stress* τ be perpendicular to the segment and greater than the *Peierls stress*.
Under its influence, the segment will *glide*, becoming simultaneously
curved, because the network nodes fix its ends (stage II). The stress,
necessary for further moving the curved segment, increases with an
increase in its curvature. The maximum curvature corresponds to a
moment when the segment attains the shape of a semicircle with the radius
$L/2$, and the corresponding stress is $\tau_{FR} = Gb/L$, where G is the *shear
modulus,* and b is the *Burgers vector.* If τ slightly exceeds τ_{FR}, the further
glide of the dislocation half-loop can evolve without an increase in τ and
it will glide further *spontaneously.* This will lead to an expansion of the
dislocation half-loop with a decrease of its curvature (stages III–V), as

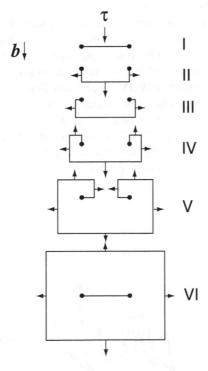

FIGURE F.3 Scheme of the Frank–Read source. A curved dislocation is substituted by a
polygonal line. Horizontal portions of the line represent edge components (both positive and
negative); vertical portions, parallel to the Burgers vector b, represent left- and right-handed
screw components. Under the influence of the shear stress τ, the opposing portions of the
same type, but the opposite sense, glide in the opposite directions, as shown by the small
arrows. As a result, the loop expands (stages II–V), its encountering portions meet and
annihilate, and a new closed loop forms, whereupon the original dislocation restores (stage VI).

well as to its transformation into a closed *glissile loop*, the transformation being accompanied by the regeneration of the initial dislocation segment (stage VI). Under the influence of the same stress, the segment can produce up to several tens of new dislocation loops. This, on the one hand, increases the *dislocation density* and, on the other, results in an increase of the *flow stress* if the loops encounter an obstacle (see *pile-up* and *Orowan mechanism*).

Frank–van der Merve growth mode Layer-by-layer growth of *heteroepitaxial films* on single-crystalline substrates; it is usually observed when the film and substrate have identical *bond* types.

Frank vector See *disclination*.

free energy Either *Gibbs' (G)* or *Helmholtz (F) free energy*, unless specified. Since in condensed *systems*, the term *PV* in the formula for Gibbs' free energy is small in comparison to the other terms, *G* in such systems is assumed approximately equal to *F*. Free energy characterizes the stability of a system; in an *equilibrium system*, it is minimal.

free enthalpy See *Gibbs' free energy*.

Frenkel pair Aggregate consisting of vacancy and self-interstitial.

full annealing *Heat treatment* of *steels* comprising a holding stage at temperatures either above A_3 (in *hypoeutectoid steels*) or between A_1 and A_{cm} (in *hypereutectoid* steels) followed by furnace cooling.

full width at half maximum (FWHM) See *x-ray line width*.

fundamental translation vector One of the vectors *a*, *b*, or *c* coinciding with the *unit cell* edge; its magnitude is equal to the edge length *a*, *b*, or *c*, respectively (see Figure F.4). Translations of the unit cell along these vectors produce the corresponding *crystal lattice*. Fundamental translation vectors are also referred to as *crystal axes*.

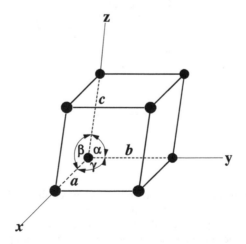

FIGURE F.4 Unit cell with fundamental translation vectors *a*, *b*, and *c*.

G

γ-Fe *Allotropic form* of iron having *FCC crystal structure* and existing between 910° and ~1400°C at atmospheric pressure.

γ′-phase In Ni *alloys*, an *intermediate phase* of *composition* $Ni_3(Ti,Al)$. The phase has the same *FCC lattice* as the *matrix*, γ *solid solution*, although the Ti and Al atoms in the γ′-phase lattice occupy cube vertices only, whereas in the γ solid solution, they occupy the *lattice sites* randomly. At a certain alloy *composition*, the *interface* between the γ′- and γ-phases is perfectly *coherent*, and the *precipitates* of γ′-phase are stable to *coarsening*.

garnet ferrite *Ferrimagnetic* oxide of *stoichiometry* $R_3Fe_5O_{12}$, where R is a trivalent rare-earth element or yttrium.

gas constant $R = 8.314$ J·mol^{-1}·K^{-1} = $8.62 \cdot 10^{-5}$ eV·K^{-1}.

general grain boundary *High-angle* grain boundary whose *disorientation* significantly differs from that of *special boundaries*. In terms of *CSL*, disorientations at general boundaries are assumed to be characterized by Σ > 25. The *atomic structure* of general boundaries is distorted, but is not *amorphous*; it is not systematically described so far. General boundary is also known as *random boundary*.

geometric coalescence Merging of two adjacent *grains* of almost the same orientation. Since their *disorientation* is small, they are separated by a *subboundary*, and their coalescence can occur without a significant rotation of their *lattices*, as in the case of *subgrain coalescence*. See also *grain coalescence*.

geometrically necessary dislocations Dislocations contributing to the *lattice* curvature inside a *grain (subgrain)* or to *disorientation* of a *low-angle boundary*. For instance, consider a grain in which there are only parallel *edge* dislocations of different *sense*. If the numbers of dislocations of the opposite sense, N^+ and N^- are the same, geometrically necessary dislocations in the grain are lacking. If, however $N^+ > N^-$ the number of geometrically necessary dislocations equals $N^+ - N^-$.

Gibbs' free energy *Free energy*

$$G = H - TS$$

69

where H and S are the enthalpy and the entropy, respectively, and T is the absolute temperature. The enthalpy

$$H = U - PV$$

where U is the internal energy, P is the pressure, and V is the volume. Gibbs' free energy is also known as free enthalpy.

Gibbs' phase rule/law Interrelation between the number of *components*, C, the number of *phases*, P, and the number of *degrees of freedom*, F, in some *equilibrium* thermodynamic *system*:

$$F = C - P + 2$$

in the case of varying temperature and pressure. If the pressure is constant,

$$F = C - P + 1$$

In a *binary* system (i.e., at $C = 2$) at a constant pressure, the independent variables can be the temperature and the *concentration* of one of the components. If $P = 1$, then $F = 2$, which means that, in a single-phase field of the corresponding *phase diagram*, both the temperature and the *composition* of the phase can be changed independently as long as the system remains single-phased. If $P = 2$, then $F = 1$, which means that, in a two-phase field, only one variable can be changed independently. In the case of the arbitrarily chosen temperature, the compositions of the phases are fixed, and in the case of the arbitrarily chosen composition of one of the phases at a given temperature, the composition of the other phase is fixed. Finally, if $P = 3$, then $F = 0$, i.e., in a three-phase field (in binary systems, it is represented by a horizontal line contacting three single-phase fields), the compositions of all the phases concerned are fixed and the phase equilibrium can take place at a constant temperature only. This is the reason why all the three-phase reactions in binary systems are termed *invariant*. In *ternary* systems ($C = 3$), invariants are reactions with four participating phases; whereas in a reaction with three participating phases, the compositions vary in the course of the reaction, and the reaction develops in a temperature range. See, e.g., *eutectic* and *peritectic reactions*.

Gibbs–Thomson equation Description of an alteration in *chemical potential*, μ, induced by a curved *interface* (free surface, *phase boundary*, or *grain boundary*):

$$\Delta\mu = 2\sigma/\rho$$

where σ and ρ are the *interfacial energy* and the radius of interface curvature, respectively. Chemical potential reduces when the interface migrates toward its center of curvature. The same effect reveals itself in the dependence of the *solubility limit* on the interface curvature:

$$\ln (c_p/c_e) \propto 2\sigma/\rho$$

where c_p is the *solubility limit* of a *solid solution* in contact with a *second-phase* particle having a curved interface with the radius of curvature ρ, and c_e is the solubility limit according to a *phase diagram* (i.e., at $\rho \to \infty$). Thus, the solubility is higher in the vicinity of a small particle than in the vicinity of a larger particle. The effect described by these equations is revealed, e.g., in *grain growth* under the influence of *capillary driving force, Ostwald ripening, spheroidization, solid-state sintering*, etc. The second equation is also known as the Thomson–Freundlich equation.

glancing angle Angle between the primary x-ray (electron) beam and the reflecting *lattice plane*. See *Bragg's law*.

glass-ceramic *Polycrystalline* ceramic material obtained by *devitrification* of a *glassy phase*; it can contain a certain amount of residual glassy phase. Devitrification is controlled in order to separate the stages of *nucleation* and growth of *crystalline phases* and to prevent a subsequent *grain growth*. All this results in an *ultra fine-grained microstructure*. Certain *nucleation agents* promote the formation of such a structure.

glass transition temperature (T_g) Temperature at which viscosity of some liquid *phase* becomes as high as $\sim10^{13}$ P, which is characteristic of *glassy phases*. The value of T_g depends on the cooling rate. The transition liquid $\to$ glass is irreversible, i.e., a glassy phase cannot transform upon heating into a liquid phase, but rather, undergoes crystallization; this proves that a glassy phase is *metastable*. See *devitrification*.

glassy phase *Amorphous metastable solid* phase obtained through cooling an inorganic liquid phase at a rate greater than *critical*. In contrast to liquids, it has high viscosity and reveals a noticeable elasticity. In *metallic glasses*, the *atomic structure* can be described by a dense random packing of hard spheres of approximately equal radii. In oxide glasses, a random network of tetrahedra, or triangles, can represent the atomic structure, with oxygen atoms at their corners and a nonoxygen atom at the centers, the tetrahedra (triangles) sharing only the corners. The atomic structure of many glassy phases can also be conceived as consisting of small (~5 nm) *crystals* or *quasicrystals* homogeneously distributed in an amorphous *matrix*. The *specific volume* of a glassy phase is greater than that of a *crystalline* solid of the same *composition*, the difference being known as excess free volume. In *crystalline ceramics*, unavoidable *impurities* can form a glassy phase that can be found as a thin (~1 nm) layer along *grain boundaries*. Glassy phase is also known as *amorphous solid* or vitreous phase. See also *glass transition temperature*.

glide Motion of *dislocations* over their *slip planes* without mass transport. See also *slip*.

glissile Able to move without mass transport.

Goss texture Preferred *grain* orientation $\{110\}\langle001\rangle$ with $\{110\}$ *lattice plane* parallel to the sheet plane and $\langle001\rangle$ *lattice direction* parallel to *RD*. A strong Goss texture with the scatter of 2–3° develops in *Fe–Si alloy* sheets because of *abnormal grain growth*. This texture significantly improves magnetic properties when an external magnetic field is applied along RD because $\langle 001 \rangle$ is an *easy magnetization direction* in these alloys (see also *magnetic texture*). The Goss component is observed in *annealing textures* of *FCC* and *BCC* metals. Goss texture is also referred to as cube-on-edge texture.

grain In *polycrystals*, an individual *crystal* of an irregular shape determined by the *nucleation* and growth conditions. It can also be referred to as crystallite.

grain aspect ratio Ratio between the mean longitudinal and the mean transverse *grain sizes*. To find the grain aspect ratio, it is necessary to choose correctly the plane of the *metallographic sample*, because, e.g., a *columnar grain* appears *equiaxed* in a transverse section and elongated in a longitudinal section.

grain boundary Interface between two *grains* of the same *phase*. Grain boundaries (GB) are identified by: a *disorientation* angle (*low-angle* and *high-angle* GB); an orientation of their plane in respect to the disorientation axis (*tilt*, *twist*, and *mixed* GB), or to the *lattices* of the adjacent grains (*symmetric-* and *asymmetric*-tilt GB); and an *atomic structure* (e.g., *special* and *general* GB). In *polycrystals*, GB form a continuous three-dimensional network wherein they are connected by lines of *triple junction* and by quaternary points. Since the grains in a polycrystal are always of different sizes, the boundaries between grains of different sizes are inevitably curved. An atomic arrangement typical of *crystal lattices* is disturbed in GB. As a result, GB are characterized by excess *grain-boundary energy*. GB can serve as *vacancy sinks*. Migrating GB can accommodate *primary dislocations* and simultaneously leave some dislocations behind them, the density of these dislocations in metals being $\sim 10^6$ cm^{-2}. At temperatures $\leq 0.4\ T_{\mathrm{m}}$, GB serve as strong obstacles to the *dislocation glide* motion (see *pile-up*), whereas at higher temperatures, they contribute to the *plastic deformation* (see *grain boundary sliding* and *diffusional creep*). GB distort the *band structure* in both *ionic* and *covalent* materials (owing to the appearance of broken or dangling bonds); the distortions can be increased by *grain-boundary segregation* of certain *impurities*. In semiconductors, GB provide *deep centers*; *coherent twin boundaries* are the least active ones in this sense.

grain-boundary allotriomorph *Crystallite* of a new *phase* nucleating at and growing over the *grain boundaries* of the parent phase. If the volume fraction of the new phase is relatively low, grain-boundary allotriomorphs form a network corresponding to the arrangement of the prior grain boundaries of the parent phase (see, e.g., *carbide network*). At a higher volume-fraction of the new phase, grain-boundary allotriomorphs are of an *equiaxed* shape.

grain-boundary character distribution Description of the distribution of *disorientation* parameters of neighboring *grains*. In non-*textured polycrystals*, the fraction of both *special* and *nearly special boundaries* is $\sim 6\%$.

Grain-boundary character distribution differs from *misorientation distribution function* in such a way that the former relates to the adjacent grains only, whereas the latter describes all the possible disorientations between all the grains in a *polycrystal*. Grain-boundary character distribution is also termed microtexture.

grain-boundary diffusion Atomic flux over the grain boundaries. It proceeds much faster than the *bulk diffusion* due to a distorted *atomic structure* of the boundaries in comparison to the *crystal lattice* (see *short-circuit diffusion path*). The *activation energy* for grain boundary self-diffusion in pure materials is ~2 times lower than that for *bulk diffusion*, excluding *special* and *low-angle* boundaries, where it is close to the activation energy for bulk diffusion. Grain-boundary diffusion is especially noticeable at temperatures $<0.4 \, T_m$.

grain-boundary dislocation *Linear defect* in a *high-angle grain boundary* whose *Burgers vector* corresponds to the *atomic structure* of the boundary, and is significantly smaller than the Burgers vector of *primary dislocation*. Grain-boundary dislocations, termed *secondary grain-boundary dislocations (SGBD)*, are usually associated with boundary ledges. There can be intrinsic and extrinsic SGBD. A system of intrinsic SGBD determines the *disorientation* at the boundary, as well as the *grain boundary orientation*. SGBD not affecting the disorientation or orientation of the boundary are called extrinsic. They occur due to the dissociation of primary dislocations trapped by the grain boundary (see *dislocation delocalization*). Extrinsic SGBD increases both the *grain-boundary energy* and *mobility* and facilitate *grain-boundary sliding*.

grain-boundary energy Excess *free energy* associated with *grain boundary*. It is connected with the distortions in the *atomic structure* of the boundary layer in comparison to the corresponding *crystal structure*. It depends on the boundary *disorientation* and, in *special boundaries*, on the boundary *orientation*. *Low-angle boundaries*, as well as the *coherent twin boundary*, are characterized by the lowest energy. The energy of *general grain boundaries* is assumed equal to ~1/3 of the energy of the free surface.

grain-boundary mobility Coefficient, M, in the formula connecting the rate of grain boundary migration, v_m, and the *driving force* for migration, Δg, at a constant temperature:

$$v_m = M\Delta g$$

Mobility depends on the boundary *atomic structure* and is affected by both the *diffusion coefficient* in the direction perpendicular to the boundary and the intensity of *grain-boundary segregation*. The temperature dependence of grain-boundary mobility is described by the *Arrhenius equation*. The observed *activation energy* for grain-boundary migration varies from the activation energy for *grain boundary self-diffusion* to that for *bulk diffusion*. The units of grain boundary mobility are m^4/Ns.

grain-boundary orientation Position of a *tilt* or *mixed grain boundary* relative to the *crystal lattices* of the grains separated by the boundary. There can be *symmetric* and *asymmetric* tilt boundaries.

grain-boundary segregation Increased *solute concentration* at a grain boundary (see *segregation*). There can be equilibrium and nonequilibrium grain-boundary segregations. The equilibrium ones are observed in cases of solute *adsorption* at the boundary. Solute concentration in equilibrium segregations depends strongly on the *solubility limit*. It can be up to 10^5 times greater than in a *solid solution* if the solubility limit is $\sim 10^{-2}$ at%, and only ~ 10 times greater in the case of a high solubility limit (~ 10 at%). The degree of grain-boundary enrichment depends on the boundary *atomic structure* and is minimum for *special grain boundaries* and maximum for *general grain boundaries*. In *ionic crystals*, equilibrium segregation is additionally affected by deviations from the charge balance due to anion/anion or cation/cation neighborhood across the boundary. Non-equilibrium segregations are not connected with *adsorption*. For instance, they can appear via migration of *vacancy*-solute atom associations to the boundaries where the vacancies annihilate and leave behind the solutes. Grain-boundary migration also results in the formation of nonequilibrium segregations because a migrating boundary sweeps out solute atoms. Both types of segregations decrease the *grain-boundary diffusivity*, which can affect the *grain-boundary mobility* and *grain-boundary sliding*. Grain-boundary segregation may either decrease or increase the grain cohesion. In semiconductors, grain-boundary segregation can neutralize dangling bounds and decrease a deteriorating effect of the boundaries.

grain-boundary sliding Displacement of *grains* with respect to one another over their boundaries at increased *homologous temperatures* and low *strain rates*, e.g., during *creep* or *superplastic* deformation. The atomic mechanism of grain-boundary sliding may comprise the *glide* motion of *grain-boundary dislocations* accompanied by grain-boundary migration.

grain-boundary strengthening Increase in the *flow stress* at temperatures below $\sim 0.4\ T_m$ due to a decrease of the *mean grain size*. *Grain boundaries* are strong obstacles to the *dislocation glide* motion at these temperatures (see *pile-up*). Thus, a reduction of the grain size increases the volume density of the obstacles, as well as the resistivity to *plastic deformation*. *Multiple slip* in the vicinity of grain boundaries, thereby supporting deformation compatibility of neighboring grains, also increases *strain hardening*. See also *Hall–Petch equation*.

grain-boundary tension Quantity equal to the *grain-boundary energy* provided the latter is independent of the *grain-boundary orientation*. This is possible for *general grain boundaries* at increased temperatures only.

grain-boundary torque Derivative of the *grain-boundary energy* with respect to an angle of deviation of the grain boundary plane from its low-energy position. Under its influence, the boundary tends to get this orientation. Grain-boundary torque is maximum for boundaries with a small Σ (see *CSL*), especially for *coherent twin boundaries*.

grain coalescence Merging of neighboring grains, which could be a mechanism of the occurrence of abnormally large grains (see *abnormal grain growth*). See also *coalescence*.

grain coarsening See *grain growth*.

grain growth *Spontaneous* process resulting in an increase of the *mean grain size* due to *grain boundary* migration during *annealing treatment* in the absence of *recrystallization* or *phase transitions*. The process evolves primarily under the effect of *capillary driving force* and is sometimes called curvature-driven grain growth. If all the boundaries are of identical energy, the boundaries of larger grains with smaller neighbors are concave and their migration results in consumption of the neighbors, which is accompanied by a decrease in the overall energy of grain boundaries per unit volume. Additionally, a decrease of the *elastic strain energy* induced by dislocations (see *strain-induced grain boundary migration*) or a reduction of the free surface energy (see *surface-energy driving force*) can affect grain growth. Various *drag forces* can inhibit grain boundary migration during grain growth. Grain growth is sometimes termed grain coarsening. See also *normal* and *abnormal grain growth*.

grain growth rate Either the growth rate of individual *grains* during the grain growth process or the rate of the process itself. The former can be described by $dD_{max}/2dt$, where D_{max} is the maximum *grain size*, and t is the duration of an *isothermal annealing;* whereas the rate of the growth process can be described by $d\bar{D}/2dt$, where $\bar{D}$ is the *mean grain size*.

grain-oriented Characterized by a *crystallographic texture*.

grain refining Reduction of the *mean grain size*, usually due to an increase of the *nucleation rate* of new grains. In *solidification*, the nucleation rate can be increased by increasing the cooling rate or by adding *inoculants* to the melt. In *primary recrystallization*, an increase of the deformation degree and the density of nucleation sites increases the nucleation rate (see, e.g., *particle-stimulated nucleation*).

grain size Linear *mean size* of grains in a single-phase material. It is usually calculated from a mean chord or a mean grain area measured on a *metallographic sample*.

grain size homogeneity See *homogeneous microstructure*.

grain size number Standard quantity characterizing the *mean grain size*. According to ASTM standards, the grain size number, N, is connected with the number of grains, n, per square inch of a *microstructure* image at *magnification* 100× as follows: $N = 1 + \log_2 n$. For instance, the mean grain diameter is ~22 μm at $N = 8$.

granular pearlite See *spheroidized pearlite*.

graphite One of the *allotropic forms* of carbon. Carbon atoms in the graphite *lattice* form flat hexagonal nets connected by *van der Waals bonds*, the interatomic bonds in the nets being *covalent*.

graphitization [annealing] Annealing of *white irons* at 800–900°C. It results in decomposition of *cementite* into *graphite* (see *temper carbon*) and *austenite*. Decomposition of cementite takes place due to its metastability relative to graphite. Since the *specific volume* of graphite is more than three times higher than that of austenite or cementite, temper carbon nucleates in pores, cracks, and other discontinuities inside *castings*.

graphitizer *Alloying element* in cast or malleable irons promoting graphite formation during *solidification* or in the course of *graphitization*.

gray [cast] iron Cast iron with *flake graphite* and a ferritic or pearlitic *matrix*, termed ferritic or pearlitic gray iron, respectively.

green In ceramic science, term relating to powder compacts and their properties before *sintering*, e.g., green compact, green density, etc.

Greninger–Troiano orientation relationship In *steels* with an increased carbon content, an orientation relationship between *martensite* (M) and *austenite* (A): $(011)_M \sim\parallel (111)_A$ and $\langle 111 \rangle_M \sim\parallel \langle 101 \rangle_A$, where $\sim\parallel$ means almost parallel. The *habit plane* in this case is $\{3\ 10\ 15\}_A$.

groove drag *Drag force* for *grain boundary* migration caused by *thermal grooves* on the free surface of thin flat objects with *columnar structure*:

$$\Delta g \cong 0.3 \gamma_{gb}/\delta$$

where γ_{gb} is the *grain-boundary energy* and δ is the object thickness.

Guinier–Preston (GP) zone In *supersaturated solid solutions*, an area of ~10 nm linear size enriched by *solute* atoms. The formation of GP zones is accompanied by the occurrence of *coherency strains*. The shape of GP zones depends on the difference in the *atomic sizes* of *solute* and *solvent*: it is nearly spherical if the difference is smaller than ~3% and disc- or rod-like if it is greater than ~5% (see also *lattice misfit*). In the latter case, the orientation of GP zones in respect to the matrix *lattice* depends on the elastic anisotropy of the latter. The occurrence of GP zones, in particular, during *natural aging* of Al alloys, is promoted by *quench-in vacancies*. The formation of GP zones results in *age hardening*, no *overaging* being observed in this case.

H

habit Shape of a *precipitate* or a *grain*, e.g., a plate-like habit, a dendritic habit, etc.

habit plane In *martensitic transformation*, a plane in the parent phase *lattice* retaining its position and remaining undistorted during the transformation. In *precipitation*, a *lattice plane* of the parent phase parallel to the flat interfacial facets of *precipitates*.

Hall–Petch equation Relationship describing an interconnection between the *flow stress* (or the *lower yield stress* in materials with the *yield point* phenomenon) σ, and the *mean grain size*, $\overline{D}$:

$$\sigma \ = \ \sigma_i + k\overline{D}^{-m}$$

where σ_i is the friction stress (it equals the flow stress in a *coarse-grained* material), k is a coefficient characterizing the *grain-boundary strengthening*, and the exponent $m = 1/2$. This effect of grain size may be connected with *pile-ups* at *grain boundaries* triggering *dislocation sources* in the adjacent grains. An increase of the grain size results in a larger number of dislocations in the pile-ups and, thus, in the onset of *slip* in the neighboring grains at a lower stress level. If the obstacles to the *dislocation glide* motion are not grain boundaries, but *subboundaries* or *twin boundaries*, the exponent m is between 1/2 and 1. In this case, $\overline{D}$ is either a mean *subgrain* size or an average distance between the twin boundaries. In *nanocrystalline* materials, $k = 0$.

hardenability Ability to form *martensite* on *steel quenching*; it can be enhanced by *alloying*.

hardening [treatment] See *quenching*.

hardness Resistance to the penetration of an object into the sample surface layer; in hardness tests, the object is called indenter. In *metallic alloys*, hardness is proportional to the *yield stress*.

Harper–Dorn creep *Steady-state creep* at low *stresses* and temperatures ≥ 0.6 T_m that evolves due to the *dislocation glide motion* controlled by *climb*. *Dislocation density* during Harper–Dorn creep does not increase, and the creep rate is described by a power law with the exponent $n = 1$ (see *power* *law creep*). *Grain boundaries* as *sources* and *sinks* for *vacancies* are not involved in the Harper–Dorn creep.

heat treatable [Material] able to increase its strength through a *heat treatment*.

heat treatment Treatment that comprises heating to a definite temperature, holding at this temperature, and cooling at a predetermined rate. The aim of heat treatment is to change the *microstructure* or *phase composition* and thus, the material properties. See, e.g., *aging treatment*, *annealing*, *full annealing*, *normalizing*, *quenching*, *tempering*, *graphitization*, *stress relief anneal*, etc. Heat treatment is also known as thermal treatment.

helical dislocation *Mixed dislocation* having changed its shape from stretched to helical due to the *climb* of its small *edge* component serving as a *vacancy sink*. The presence of helical dislocations is considered an indication of previous vacancy supersaturation.

Helmholtz free energy *Free energy*

$$F = U - TS$$

where U and S are the internal energy and the entropy, respectively, and T is the absolute temperature. Compare with *Gibbs' free energy*.

heteroepitaxial film *Epitaxial film* whose *composition* or *crystal structure* differs from those of the substrate. The *interface* between the film and the substrate can be either *coherent* or *partially coherent*, depending on the *lattice misfit* (see *lattice-matched* and *lattice-mismatched epitaxy*). In thin heteroepitaxial films, the interface can be coherent despite a large lattice misfit; in this case, noticeable *elastic deformations* (known as *coherency strains*) occur in the film, which is referred to as *strained-layer epitaxy*. In thicker films, the coherency strains can be partly compensated by *epitaxial (misfit) dislocations*.

heterogeneous microstructure In single-*phase* materials, a *microstructure* characterized by a *duplex grain size*. It is usually observed when *abnormal grain growth* is not completed or if *primary recrystallization* evolves differently in different zones of an article. In materials with several *microconstituents*, the microstructure is considered heterogeneous when there is a nonuniform distribution of the microconstituents over the article's cross-section.

heterogeneous nucleation Occurrence of new phase *nuclei* at *lattice imperfections* in a parent *phase* or on *crystals* of a foreign matter. The imperfections may be individual *dislocations* and their agglomerations, *stacking faults*, *subboundaries*, *grain boundaries*, and *phase boundaries*. This is explained by a decrease of the energy required for creating a new *interface* between the nucleus and the parent phase, as well as the *elastic strain energy* associated with the *specific volume* difference between the new phase and the parent one (see *critical nucleus*). All the nuclei occurring inside the grains of the parent phase have low-energy *coherent* or *partially coherent* interfaces, whereas those at the grain boundaries have either a coherent (or partially coherent) interface with only one of the adjacent matrix grains, or *incoherent* interfaces with all of them. When the nucleus composition differs from that of the parent phase, heterogeneous nucleation is additionally promoted by *solute segregation* at the previously mentioned defects. Small crystals of a certain foreign matter can promote heterogeneous nucleation if they serve as substrates for the *epitaxial* growth of nuclei, which decreases the overall energy of the new interface (see *inoculant* and *nucleation agent*).

heterogeneous system Thermodynamic *system* consisting of more than one *phase*.

heterojunction *Coherent or partially coherent* flat *phase boundary* between *amorphous* and *crystalline* phases or between two different crystalline phases, if at least one of them is semiconducting. The *segregation* of certain *impurities* can neutralize the trapping effect of heterojunctions.

heterophase Consisting of many *phases*.

heteropolar bond See *ionic bond*.

heterostructure See *superstructure* (in microelectronics).

hexagonal close-packed (HCP) structure One of the two most densely packed *crystal structures* (along with *FCC*) corresponding to the *hexagonal system* (see Figure H.1). It has *axial ratio* $c/a = 1.633$, *coordination number* 12, *atomic packing factor* 0.74, and the *close-packed planes* {0001} and *close-packed directions* $\langle 11\bar{2}0 \rangle$. The radius of the *tetrahedral void* equals $0.225R$, and that of the *octahedral void* is $0.415R$, where R is the *atomic radius*. In HCP structure, the atom centers of the second densely packed layer {0001} are arranged in the B hollows of the first layer (see Figure H.2), and the atom centers of the third layer are arranged in the hollows of the second layer above the A centers of the first layer. Thus, the stacking sequence of the atom layers {0001} in HCP structure can be described as ...*ABABA*... (or ...*ACACA*...). See *Miller–Bravais indices*.

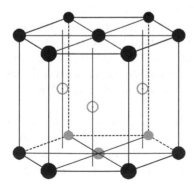

FIGURE H.1 Unit cell of HCP lattice.

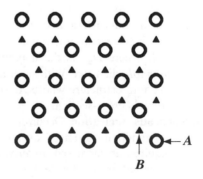

FIGURE H.2 Sequence of close-packed planes {0001} in HCP lattice.

hexagonal ferrite *Ferrimagnetic* oxide of *stoichiometry* $MFe_{12}O_{19}$, where M is divalent Ba, Sr, or Pb. The name has its origin in the fact that O^{2-} ions form a hexagonal *sublattice* with M^{2+} ions occupying some oxygen sites; the Fe^{3+} ions form a second sublattice, arranging in certain *tetragonal* and *octahedral voids* of the first one.

hexagonal system *Crystal system* whose *unit cell* is characterized by the following *lattice parameters*: $a = b \neq c$, $\alpha = \beta = 90°$, $\gamma = 120°$.

high-angle grain boundary Grain boundary (GB) whose *disorientation* angle exceeds $\sim 15°$. The *atomic structure* of high-angle GBs is ordered to some extent, but cannot be described by a two-dimensional array of *primary dislocations* as *low-angle boundaries* can. The simplest description of the GB atomic structure provides a *CSL* model. A more complicated description is given by models of structural units, according to which, definitely ordered groups of atoms (known as structural units) are periodically arranged over the GB plane. The number and arrangement of atoms in structural units differ in GBs of different *disorientations* and *orientations*. The thickness of *random* high-angle GBs is 2–3 *lattice constants*. High-angle GB is also termed large-angle grain boundary. See also *special grain boundary*.

high-resolution transmission electron microscope (HRTEM) *TEM* supplying an image of *crystal lattices* (i.e., it achieves the lateral *resolution* of ~ 0.2 nm).

high-temperature thermo-mechanical treatment Treatment consisting of *hot deformation* of a high-temperature *phase* immediately followed by *quenching* to temperatures below M_f. The deformation induces *dynamic recrystallization* in the phase, which leads to *grain refinement*. This, in turn, results in a decreased size of the *martensite* crystallites appearing upon quenching.

high-voltage electron microscope (HVEM) *TEM* with accelerating voltage ~ 1 MV; it is used for studying foils of a greater thickness than a conventional TEM.

homoepitaxial film *Film* made of the same material as the substrate. Owing to *epitaxy*, its orientation is the same as that of the substrate.

homogeneous microstructure In a single-*phase* material, a microstructure in which the range of *grain sizes* can be described by the ratio $D_{max}/D_M \cong$ 3–4, where D_{max} is the maximum grain diameter, and D_M is the *most probable* diameter. With several *microconstituents*, the same term relates to a microstructure whose constituents are uniformly distributed over the article's cross-section.

homogeneous nucleation Occurrence of new *phase nuclei* when nucleation sites are randomly arranged over the parent phase. This kind of nucleation requires an increased *undercooling* (see *critical nucleus*) and is rarely observed in solid-state transformations. Compare with *heterogeneous nucleation*.

homogeneous system Thermodynamic *system* consisting of only one *phase*.

homogenizing [anneal]/homogenization *Heat treatment* consisting of a prolonged holding at increased temperatures with the aim of decreasing chemical inhomogeneity due to *coring*.

homologous temperature Ratio T/T_m, T being the absolute temperature concerned, and T_m being the *melting point* of the *base component* (in K).

homopolar bond See *covalent bond*.

Hooke's law Existence of direct proportionality between *stress* and *strain* at small (<0.01%) strains, with a proportionality coefficient known as *elastic modulus*. See *Young's modulus*, *shear modulus*, and *bulk modulus*.

host atom *Solvent* atom.

hot deformation Procedure of *plastic deformation* above the *recrystallization temperature*.

hot isostatic pressing (HIP) See *hot pressing*.

hot pressing *Solid-state sintering* at temperatures above ~0.6 T_m with a simultaneous application of pressure. This permits a decrease in *firing* temperature, which prevents the possible development of *abnormal grain growth*. If the pressure is hydrostatic, the process is called hot isostatic pressing.

hot worked Subjected to *hot deformation*.

hot-stage microscope *Optical* or *electron microscope* with an attachment for *in situ observations* in the course of specimen heating.

Hume–Rothery phase *Intermediate phase* of a varying *composition* characterized by an approximately constant *electron concentration* that can be 1.5, 1.62, or 1.75 for different *crystal structures*. It is also referred to as electron compound/phase.

Hume–Rothery rules Empirically found conditions for unlimited *solubility* of *substitutional solutes* in a certain *solvent*: a difference in *atomic radii* of the *components* <15%; identity of the *crystal structure* and chemical nature of the solute and solvent; and a small difference in the *electronegativity* of the components.

hydrostatic pressure Three-axial compression wherein all three principal *normal stresses* are equal and *shear stresses* are lacking.

hypereutectic [Alloy] possessing a *composition* to the right of an *eutectic point* (in a *binary phase diagram*) and undergoing the *eutectic reaction*.

hypereutectoid [Alloy] possessing a *composition* to the right of an *eutectoid point* (in a *binary phase diagram*) and undergoing the *eutectoid reaction*.

hypoeutectic [Alloy] possessing a *composition* to the left of an *eutectic point* (in a *binary phase diagram*) and undergoing the *eutectic reaction*.

hypoeutectoid [Alloy] possessing a *composition* to the left of an *eutectoid point* (in a *binary phase diagram*) and undergoing the *eutectoid reaction*.

I

ideal orientation See *texture component.*

immersion objective/lens In *optical microscopes*, an objective with a *numerical aperture* $A_N > 1.0$ (up to ~1.30). It works with a special medium between the lens and an object whose refraction index exceeds 1.0.

imperfect dislocation See *partial dislocation.*

impurity Incidentally present substance or chemical element, unlike *alloying element*. In semiconductors, impurity frequently means the same as *dopant.*

impurity cloud See *Cottrell* and *Suzuki atmospheres.*

impurity drag Inhibition of *grain boundary* migration by equilibrium *grain-boundary segregations*. Since the segregations decrease the *grain-boundary energy*, they reduce the *capillary driving force*, thus causing a *drag force*. At the same time, impurity drag is most often used in the sense that the segregated *impurity* reduces the effective *mobility* of grain boundaries because the *diffusivity* of impurity atoms differs from that of the *host* atoms. Impurity drag is also called solute drag.

incoherent interface *Phase boundary* in which there is no coincidence of the *lattice points* of neighboring lattices, in contrast to *coherent* or *partially coherent interfaces.*

incoherent precipitate/particle *Second phase* precipitate whose *interface* with the *matrix phase* is *incoherent*. Incoherent precipitates have little to no *orientation relationship* with the *matrix.*

incoherent twin boundary Twin boundary whose plane does not coincide with the twinning plane (see *twin*). A boundary of this kind is always joined to either a *coherent twin boundary* or the boundaries of the *twinned grain.* The energy and mobility of an incoherent twin boundary are rather close to those of *general high-angle grain boundaries*, in contrast to a *coherent twin boundary.*

incubation period In materials science, the time duration (at a constant temperature) necessary for the first stable *nuclei* of a new *phase* to occur. The incubation period found experimentally is often greater than the true incubation period, due to an insufficient sensitivity of investigation techniques used. Incubation period is sometimes called induction period.

indirect replica See *replica.*

induction period See *incubation period.*

inelastic scattering Interaction of x-rays (electrons) with a *crystalline* material accompanied by changes in the wavelength (energy) of scattered radiation. For instance, *secondary electrons*, including *Auger-electrons*, arise because of an inelastic electron scattering. *X-ray fluorescence* and *Compton scattering* result from an inelastic x-ray scattering.

ingot In metallurgy, a product obtained through the *solidification* of liquid metal in a mold. The ingot *macrostructure* consists of three zones: a chill zone of *equiaxed*, relatively small *grains* close to the mold wall, a *columnar zone*, and a central equiaxed zone. A *shrinkage* cavity develops in the upper-central part of a metallic ingot due to a decrease of the *specific volume* during solidification. In ingots obtained by *directional solidification*, there is only a columnar zone. *Single crystals* can be also obtained by crystallization using *seed crystals.*

inhomogeneous microstructure In single-*phase* materials, a *microstructure* characterized by a range of *grain sizes* where the ratio $D_{max}/D_M > 3–4$ (D_{max} is the maximum grain diameter, and D_M is the *most probable* diameter). Inhomogeneous microstructure appears either after *primary recrystallization*, if the preceding deformation or the spatial distribution of *recrystallization nuclei* is inhomogeneous, or when *abnormal grain growth* is incomplete (see *duplex grain size*). In multiphase materials, the same term relates to the microstructure whose constituents are nonuniformly distributed over the article's cross-section.

inoculant Auxiliary substance causing *grain refinement* of a *phase* appearing during *solidification*. *Crystals* containing inoculant appear before the solidification of an *alloy* starts and serve as substrates for the *epitaxial* growth of the solid-phase *nuclei*. Since in this case, the *interface* between the nucleus and the substrate is *coherent* or *partially coherent*, the nucleus' *interfacial energy* is decreased, which results in an increased rate of *heterogeneous nucleation* and thus grain refinement. Substances facilitating heterogeneous nucleation of *crystalline* phases in a *glassy phase* are called *nucleation agents* (see *glass-ceramic*).

in situ **observation** Direct observation of a dynamic process.

instrumental [x-ray] line broadening Increase of an *x-ray line width* due to: a wavelength range in primary radiation; an incorrect adjustment of the sample; a varying thickness of diffracting layer; and a vertical divergence of the primary beam. Any *crystal monochromator* also increases instrumental broadening.

integral [x-ray] line width Characteristic of an *x-ray diffraction line* found by dividing the *integrated intensity* of the line by its height (both of the quantities are to be measured from the *background* level). See *x-ray line intensity.*

integrated [x-ray] line intensity Area under a peak on a *diffractogram* measured from the background level. Integrated intensity is used instead of the *line intensity* because the deviations from the unique wavelength and from x-ray parallelism in the primary beam can affect the latter. Integrated intensity is proportional to the volume fraction of the diffracting *phase*.

interatomic spacing In *metallic* and *covalent crystals*, it is assumed equal to half the distance between the two closest similar atoms (i.e., 2 *atomic radii*); in *ionic crystals*, it equals half the sum of the anion and cation radii.

intercritical heat treatment Heat treatment comprising a heating stage at temperatures inside an *intercritical range*.

intercritical range Temperature range between the boundaries of a two-*phase* field in the corresponding *phase diagram*.

intercrystalline Evolving over *grain boundaries* or *phase boundaries*. Intergranular is a synonym for intercrystalline.

interdiffusion See *chemical diffusion*.

interface Normally, a *phase boundary* between *grains* of different solid *phases*. However, the same term can be applied to *grain boundaries* or to the boundary between a multiphase *colony* and the parent phase.

interface-controlled [Process whose rate is] determined by the reaction rate at an *interface* or by the interface geometry (see, e.g., *diffusional transformation* and *Widmannstätten ferrite*) and is independent of the *diffusion* rate.

interfacial energy Excess *free energy* of a *phase boundary* resulting from the *lattice misfit* and the difference in the chemical nature of the neighboring atoms across the interface (if the phases are of different *compositions*). The first constituent of the interfacial energy is the lowest for *coherent* interfaces and the highest for *incoherent* ones. Interfacial energy can be called *surface tension,* provided the energy does not depend on the interface structure as in incoherent interfaces.

intergranular See *intercrystalline*.

interlamellar spacing In materials science, the mean distance between lamellae in a *colony.*

intermediate phase *Solid phase* whose field in the *phase diagram* does not include pure *components* of the *system* concerned. Intermediate phases can be of a constant or a varying *composition* (the latter are usually observed in *metallic* systems). See *Hume–Rothery phase, intermetallic compound, interstitial phase, Laves phase, carbide.*

intermetallic compound *Intermediate phase* formed by metallic *components* and having a realatively constant *composition*.

internal friction Technique for studying *anelasticity* by imposing free *strain* oscillations on a thin specimen. Internal friction

$$Q^{-1} \cong \delta/\pi$$

at low δ, where

$$\delta = \ln(A_n/A_{n+1})$$

is known as logarithmic decrement, and A_n and A_{n+1} are the amplitudes of two consequent oscillations. The maximum δ is attained at $\omega\tau_R = 1$, where ω is the oscillation frequency, and τ_R is the *relaxation time*. Since τ_R is a temperature-dependent constant, measurements of Q^{-1} at a constant ω yield the $Q^{-1}(T)$ dependence, with a maximum referred to as internal friction peak (see *Bordoni peak, Kê peak, Köster peak, Snoek peak,* and *Zener peak*). Internal friction peaks observed in *glassy phases* are frequently connected with *diffusion* of certain cations. The temperature dependence of τ_R, and thus the *activation energy* of the corresponding dissipation process, can be gained from studying the $Q^{-1}(T)$ dependence at different temperatures. In some cases, an amplitude-dependent internal friction is observed (see *Köster effect*). Internal friction in *crystalline ceramics* and *glass-ceramics* increases on approaching the *glass transition temperature*.

internal oxidation *Nucleation* and growth of oxide particles inside a *solid solution* containing readily oxidizable *solutes*, as, e.g., aluminum or silicon in copper. The oxides occur during *annealing* due to the diffusion of oxygen into the specimen from the environment. Internal oxidation is one of the methods for producing *ODS alloys*.

internal stresses See *residual stresses*.

interphase precipitation In low-*alloy steels*, straight parallel rows of small *carbonitride* particles inside *ferrite grains*. The rows occur in the course of the growth of platelet-like ferrite grains that proceed by the intermittent ledge motion over their flat *interfaces* with *austenite*.

interplanar spacing Distance between a pair of the closest parallel *lattice planes* measured along the normal to the planes. The *close-packed planes* have a maximum interplanar spacing.

interstice See *lattice void*.

interstitial See *interstitial atom*.

interstitial compound See *interstitial phase*.

interstitial [foreign] atom *Solute* atom occupying an *interstice* in a *host* lattice. See *point defects*.

interstitialcy See *self-interstitial*.

interstitial [mechanism of] diffusion Diffusion by atomic jumps from one *interstice* to an adjacent one. Only *interstitial solute* atoms can diffuse in such a way because their *atomic radius* is close to that of an interstice. This mechanism is characterized by smaller *activation energy* than that found in *vacancy mechanism*.

interstitial phase *Intermediate phase* of a varying chemical *composition* close to the *stoichiometry* MeX or Me_2X, where Me is a transition metal, and X is a nonmetal (usually hydrogen, carbon, or nitrogen) whose *atomic radius* is between 0.41 and 0.59 of the radius of the Me atoms. The *crystal structure* of interstitial phases can be described as consisting of two *sublattices*, one formed by the Me atoms, and the other by the X atoms. The former can be *FCC, BCC,* or *HCP*, whereas the sites of the latter lie in certain *octahedral voids* of the former. Interstitial phase is also known as interstitial compound.

interstitial solid solution Solid solution wherein *solute* atoms are arranged in *lattice voids* of the *host* lattice. This is only possible when the *atomic size* of the solute atoms is smaller than that of the host atoms, as, e.g., for carbon and nitrogen in iron, or oxygen in titanium. Interstitial solutes usually bring about severe *static lattice distortions* and, because of this, an increased *solid-solution strengthening*.

intracrystalline See *transcrystalline*.

intragranular See *transcrystalline*.

intrinsic diffusion coefficient *Diffusivity* of atoms of a *component* taking part in *chemical diffusion*.

intrinsic diffusivity See *chemical diffusion*.

intrinsic grain-boundary dislocation *Grain-boundary dislocation* defining the *atomic structure* of a boundary with certain *disorientation* and *orientation*.

intrinsic stacking fault Disturbance in the stacking sequence of the *close-packed planes* caused by the lack of a part of one plane, e.g., due to the collapse of a *vacancy* disc lying in the plane. For instance, in *FCC structure* with a perfect stacking sequence ...ABCABCABC..., removing a plane **B** results in a sequence ...ABCACABCA... with the layers AC arranged erroneously (their stacking sequence is typical of *HCP structure*). Because of this distorted stacking sequence, intrinsic stacking fault is characterized by a certain excess energy. Since it is geometrically identical to a *twin* one layer in thickness (the layer sequence in a twin is ...ABCABACBA...), its energy is two times greater than the energy of the *coherent twin boundary*.

intrinsic [x-ray] line broadening Increase of *x-ray line width* due to the small size (<0.5 µm) of mosaic blocks (see *mosaic structure*), nonuniform *microstrain*, and *stacking faults*, as well as *coring*. There are several methods for deriving both the mean block size and the average microstrain from the magnitude of intrinsic broadening (see, e.g., *Warren–Averbach method*).

invariant reaction In thermodynamics, a *first-order phase transformation* wherein the number of the *degrees of freedom* equals zero (see *Gibbs' phase rule*); i.e., temperature, pressure, *phase constituents*, and their *compositions* remain constant in the course of the reaction.

inverse pole figure Distribution of certain sample directions (e.g., *RD*, *TD*, and *ND* in a rolled sheet) with respect to specific *lattice directions*. Inverse pole figures are presented in the *standard stereographic triangle*.

ion channeling Technique using a collimated beam of high-energy (~1 MeV) ions directed along a highly symmetric *lattice direction* in a *crystal*. The *energy spectrum* of the scattered ions can give data on: the *atomic structure* of the crystal surface; the presence and *atomic structure* of *adsorption* layers; the atomic structure of the *interface* between the adsorption layer and the crystal; the arrangement of *impurity atoms* in a *crystal lattice;* an in-depth distribution of *crystal defects;* the thickness of the *amorphous* (or *polycrystalline*) layer; and on crystal orientation.

ion etching Technique for revealing *microstructure* by *sputtering* the surface of polished *metallographic samples*. The ion-etching rate depends (along with the ion energy and flux density) on the bond type (i.e., *metallic, covalent*, or *ionic*) as well as on the masses of atoms in different *phases* of the sample. This technique is most frequently used for preparing *SEM* samples.

ionic bond Bond type between atoms of typical metallic and nonmetallic *components*. The former lose their valence electrons, thereby becoming cations, while the latter acquire these electrons, thus becoming anions. Electrostatic attraction between cations and anions is compensated by the repulsion between the ions of the same sign. This bond also referred to as heteropolar.

ionic crystal Electrically neutral crystal whose ions are held at the *lattice sites* by *ionic bonds*; in some cases, the interatomic bonds in the crystals can be mixed, i.e., partially ionic and partially *covalent* (see *electronegativity*). Since ionic bonds are *isotropic*, the ions arrange so that the greatest possible amount of ions of the same sign surround an ion of the opposite sign. The number and the arrangement of the surrounding ions (i.e., the shape of the *coordination polyhedron*) are governed, along with electrostatic interaction, by a geometric factor depending on the ratio of the ion radii. *Crystal structure* of the majority of ionic crystals is composed of an *FCC* or an *HCP sublattice* formed by larger ions and sublattices of smaller ions occupying certain voids in the first sublattice (see, e.g., *CsCl, NaCl*, and *CaF2 crystal structures*). The stability of ionic crystal structures is supposedly highest if the coordination polyhedra share vertices. Since ionic crystals are electrically neutral, any *lattice defects* disturb both the charge balance and the *band structure*. To compensate for the disturbances, they form associates with lattice defects of the opposite sign (e.g., *Frenkel* or *Schottky pairs*) or with electrons (see, e.g., *color center*).

ionic radius Conventional ion size in a *lattice* of *ionic crystals* (see *atomic radius*). Ionic radius is found by calculations based on the experimentally measured radius of the O^{2-} ion. Tabulated values of ionic radii relate to a *coordination number* equal to 6.

ion implantation Procedure for changing chemical *composition* and, inevitably, *substructure* and, sometimes, *atomic structure* of the surface layers by bombarding with high-velocity positive ions. The ions intruding into the surface layer of thickness ~100 nm produce a rich variety of *point defects*, such as *vacancies, self-interstitials, foreign atoms*, and sometimes an *amorphous layer* (see *irradiation damage*). The lattice distortions are removed by a subsequent *annealing*; the treatment can restore even the initial *single-crystalline* structure.

irradiation damage Effect of irradiation with high-energy (usually ~1 MeV) ions or thermal neutrons, revealing itself in the production of nonequilibrium *vacancies, self-interstitials, crowdions*, and *displacement cascades* in a solid body. All these imperfections increase drastically the *diffusion* rate at low temperatures and thus promote the *precipitation* in *supersaturated solid solutions* (see *irradiation hardening*). Irradiation damage is also called radiation damage.

irradiation defects Primarily nonequilibrium *vacancies* and *self-interstitials,* as well as *crowdions* and *displacement cascades*. The *point defects*, primarily self-interstitials, tend to form clusters, transforming eventually into *prismatic dislocation loops*. See *irradiation damage*.

irradiation growth In irradiated materials with noncubic *crystal lattice*, dimensional changes caused by a preferable arrangement of *point defect* clusters on certain *lattice planes*.

irradiation hardening Strengthening of irradiated materials due to clusters of *vacancies* or *self-interstitials* produced by irradiation, primarily due to irradiation-induced *precipitation*, provided the *matrix phase* was *supersaturated* before irradiation. A distinctive feature of irradiation hardening is that it proceeds at relatively low temperatures ($<0.2\ T_\mathrm{m}$), because of a high *diffusion* rate caused by an increased *concentration* of *point defects*.

irradiation-induced creep In irradiated materials, *creep* deformation at temperatures below $\sim0.3\ T_\mathrm{m}$ whose rate is almost independent of temperature. The latter is a result of an increased *concentration* of irradiation-produced *vacancies* and *self-interstitials*; they enhance the *climb* of *dislocations* and thus, the creep rate.

irreversible See *reversibility*.

island film See *Vollmer–Weber growth mode*.

isochronal annealing Series of *annealing treatments* of the same duration, but at different temperatures.

isoforming *Thermo-mechanical treatment* of *hypoeutectoid steels* comprising a *hot deformation* stage at temperatures of *pearlitic range*. The deformation is started when the *pearlitic reaction* does not yet commence and is finished after the reaction is complete. As a result, instead of *proeutectoid ferrite* and *pearlite colonies* after annealing, the *microstructure* after isoforming consists of ferrite with a clearly developed *subgrain structure* (with the mean subgrain size of ~0.5 μm) and *spheroidized cementite* particles.

isomorphism Presence of mutually soluble *components* or *phases*.

isomorphous phases Solid *phases* having the same *crystal lattices*, but different *lattice constants* and different *composition*.

isomorphous system Thermodynamic *system* whose *components* are characterized by an unlimited solubility in both the liquid and solid states.

isothermal Evolving at a constant temperature.

isothermal transformation *Phase transition* that can evolve and finish at some fixed temperature. The temperature is always lower than the temperature of the corresponding transformation in a *phase diagram* because a certain *undercooling* is necessary for the *nucleation* commencement. See *diffusional phase transformation* and *critical nucleus*.

isothermal transformation diagram See *TTT (time-temperature-transformation) diagram*.

isotropic Possessing *mechanical* or *physical properties* independent of or slightly dependent on the direction in a sample. The observed isotropy in non-*textured polycrystals* is macroscopic because it results from averaging the anisotropic properties of many *crystallites* over the specimen volume. Thus, strictly speaking, nontextured polycrystals are quasi-isotropic.

J

jog Step in a *dislocation* line perpendicular to its *slip plane*. Jogs of the height equal to one *interplanar spacing* (called elementary jogs) can occur either as a result of the dislocation intersection with a *forest dislocation* or due to the dislocation *climb*. Jogs of the edge type on *screw dislocations* can also form during *double cross-slip*, the jogs being several interplanar spacings high (called multiple jogs). The edge-type jogs on screw dislocations impede their *glide* because the jogs can only climb.

Johnson–Mehl–Kolmogorov equation Description of transformation *kinetics*, assuming that: *nucleation* is *homogeneous*; the nucleation rate, $\dot{N}$, is independent of time; and the *linear growth rate* of new *crystallites*, G, is constant and isotropic. In this case, the kinetic equation is:

$$V/V_0 = 1 - \exp(-\pi \dot{N} G^3 t^4 / 3)$$

where V and V_0 are the transformed and initial volumes, respectively, and t is the transformation time. Compare with *Avrami equation*.

K

Kelvin's tetrakaidecahedron See *tetrakaidecahedron*.

Kê peak/relaxation *Internal friction* peak induced by a viscous behaviour of *grain boundaries* connected with *grain-boundary sliding*.

Kerr microscopy Technique for observing *domain structures* by means of *polarized-light microscopy*. *Magnetic domains* are visible because the interaction of polarized light with the magnetized matter rotates the polarization plane of reflected light by an angle dependent upon the magnetization direction.

Kikuchi lines Nearly straight, black lines and white lines in *electron diffraction patterns* received from relatively thick *foils*. They result from diffraction of *nonelastically scattered* electrons. Kikuchi lines are used for a precise determination (with an accuracy better than 1°) of *lattice* orientation.

kinetics [of transformation] Time dependence of the volume fraction transformed. The term relates not only to *phase transitions*, but also to *recrystallization*. Kinetics of *thermally activated* reactions are usually described either by the *Johnson–Mehl–Kolmogorov equation* or the *Avrami equation*.

kink A step of one *interatomic spacing* length in a *dislocation*. Unlike *jogs*, it lies in the *slip plane* of the dislocation. Kinks can occur by *thermal activation* as, e.g., double kinks connecting two parallel segments of the same dislocation lying on both sides of the *Peierls barrier*. Kinks can easily *glide* along with the dislocation.

kink band Part of a *plastically deformed grain*. The lattice inside the band is disoriented relative to the lattice of the grain parts outside it. In contrast to *deformation bands*, the *disorientation* inside kink bands is rather small, if any. Kink bands are frequently observed in *HCP* metals subjected to compression.

Kirkendall effect Manifestation of the difference in diffusion rates of the atoms of different components in the case of *vacancy mechanism*, i.e., in *substitutional solid solutions*. An *annealing* of a sample made of two closely contacting parts of different *compositions* results in the following. First, the interface between these parts shifts into the part with the higher *intrinsic diffusivity* (just this is known as the Kirkendall effect). Second, since the atomic flux is equivalent to the vacancy flux in the opposite direction, the excess vacancies are accumulated in the extending part, which leads to the formation of voids known as diffusion porosity. See *chemical diffusion*.

Kossel line pattern Curved black or white lines in an x-ray diffraction pattern. They result from the interaction of a strongly divergent *monochromatic* x-ray beam with a *crystal lattice* and are produced by the x-rays diffracting on the *lattice planes* whose arrangement corresponds to *Bragg's law*. These lines are used for a precise determination of *lattice constants*. Strictly speaking, the described pattern should be named pseudo-Kossel, because the true Kossel pattern occurs when an x-ray source is arranged inside a specimen.

Köster effect Amplitude dependence of *internal friction* induced by an irreversible motion of *dislocation* segments released from *solute* atoms and other obstacles. This effect has a hysteresis nature.

Köster peak/relaxation *Internal friction* observed in *cold-worked* and *aged* *BCC* transition metals due to an irreversible motion of *dislocation* segments pinned by *solute* atoms, the motion rate being controlled by the *solute* diffusion. Köster peak is also known as Snoek–Köster peak. This should not be confused with *Köster effect*.

Kurdjumov–Sachs orientation relationship Orientation of *ferrite* (F) with respect to the parent *austenite* (A): $(110)_F \parallel (111)_A$, $\langle 111 \rangle_F \parallel \langle 110 \rangle_A$. In this case, the *habit plane* is $\{111\}_A$ as, e.g., in *Widmannstätten ferrite*. The same orientation relationship is observed between *martensite* and austenite in steels with a relatively low carbon content, and the habit plane in this case is $\{111\}_A$ or $\{225\}_A$, depending on the carbon content in the martensite.

L

laminar slip Stage of *plastic deformation* of a *single crystal* in which only one *slip system* is active. See also *easy glide*.

Lankford coefficient See *r̄-value*.

large-angle grain boundary See *high-angle grain boundary*.

Larson–Miller parameter Empirically found quantity, *P*, used for the extrapolation of the creep-rupture data obtained on a definite material at a certain temperature, *T*:

$$P = cT (A + B \ln t)$$

where *c*, *A*, and *B* are empirical coefficients, and *t* is the creep-rupture life. The extrapolation is valid only for the same material and an identical stress state.

latent hardening Decreased deformation, if any, on one or more *slip systems* having the same *Schmid factor* as the *active slip system*. Latent hardening results from the inhibition of the dislocation *glide* motion, as well as from the blocking of the *dislocation sources* on the systems with low deformation, both caused by a stress field produced by *dislocations* on the active slip system.

lath martensite Product of *martensitic transformation* in low- to medium-carbon *steels* with a relatively high M_s temperature. The martensite *crystallites* have an appearance of tightly arranged, thin (<1 μm) laths separated by *low-angle boundaries*. The laths are characterized by the *habit plane* $\{111\}_A$, whereas their *lattice* is oriented with respect to the *austenite* according to the *Kurdjumov–Sachs orientation relationship*. The laths form packets with jugged boundaries. The *substructure* of lath martensite is characterized by high density of *dislocation tangles*. Lath martensite is also called packet, blocky, or massive martensite.

lattice See *crystal lattice*.

lattice basis Group of atoms belonging to a *lattice point*. Translations of lattice basis along 3 *crystal axes* build the corresponding *crystal structure*. If the group consists of one atom, the basis is denoted by the coordinates of the atoms at the origin and inside the unit cell. For instance, the basis for *BCC structure* is (0 0 0; 1/2 1/2 1/2) and for *FCC structure* (0 0 0; 1/2

1/2 0; 0 1/2 1/2; 1/2 0 1/2). If the atoms are of different types, their coordinates in the basis notation are given separately. For instance, the basis for *CsCl structure* is: (Cs: 0 0 0; Cl: 1/2 1/2 1/2). For the notation of atom coordinates, see *lattice point*.

lattice constant Length of one of the *unit cell* edges.

lattice diffusion See *bulk diffusion*.

lattice direction In crystallography, a vector starting at a *lattice point* (chosen as the origin) and ending at a point with coordinates A, B, and C. The quantities A/a, B/b, and C/c can be fractional or integer (a, b, and c are *lattice constants*), but they must be converted into a set of the smallest integers: u, v, and w. In a three-number notation, indices of the lattice direction, as well as those of all the directions parallel to it, are the smallest integers in square brackets: [uvw]. For instance, [100] denotes the x-axis in its positive direction, and [$\bar{1}$00] denotes the same axis in the negative direction. There can be several lattice directions with the same indices arranged differently and having different signs (see *form*). In a four-number notation for *hexagonal systems*, the indices of [UVTW] lattice direction can be found from Miller indices [uvw] as follows: $U = 2u - v$, $V = 2v - u$, $T = -(u + v)$, and $W = 3w$. The obtained numbers U, V, T, and W are to be reduced to the smallest integers.

lattice-matched [epitaxial] film Thin *heteroepitaxial film* whose *lattice misfit* with the substrate is small. Because of this, the lattice of the film and that of the substrate coincide almost perfectly at the *interface* plane, making the interface *coherent*.

lattice misfit Mismatch in the arrangement of *lattice sites* of different *phases* at a flat *phase boundary*. If the in-plane site patterns are the same, the lattice misfit can be defined by the parameter:

$$\delta = (a_\alpha - a_\beta)/a_\alpha$$

where a_α and a_β are in-plane *interatomic spacings* in α- and β-phase lattices, respectively (it should be stressed that a_α and a_β do not have to be *lattice constants*). In the case of *second-phase precipitates*, misfits smaller than 1–3% can be compensated by *coherency strains*, either in the precipitate or in the *matrix*. In thin *heteroepitaxial films*, coherency strains can compensate a lattice misfit up to ~5%. A greater misfit is compensated by *misfit dislocations*. The distance between the dislocations, p, can be estimated from the formula:

$$p = b/\delta$$

where b is the *Burgers vector* of the misfit dislocations.

lattice-mismatched [epitaxial] film Thin *heteroepitaxial film* whose *lattice misfit* with the substrate is compensated by *coherency strains* (known as *strained-layer epitaxy*) and *misfit dislocations*. In the latter case, the interface film-substrate is *partially coherent*.

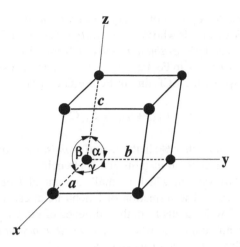

FIGURE L.1 Unit cell and its parameters.

lattice parameters Lengths of *unit cell* edges and angles between them (see Figure L.1). Frequently, the term lattice parameter (or *lattice constant*) relates to the edge length only, whereas the angles between the edges are termed *axial angles*. See also *crystal system*.

lattice plane Flat net of *lattice points*. Various lattice planes differ in the density of lattice points per unit area, as well as in *interplanar spacing*. For plane designation, see *Miller* and *Miller–Bravais indices*. Lattice plane is also referred to as net plane.

lattice point Geometric point representing an atom or a group of atoms in a *crystal lattice*; it is also referred to as lattice site. The coordinates of a lattice point are defined by the point projections, *A*, *B*, and *C*, on *crystal axes* divided by the corresponding *lattice constants*: *A/a*, *B/b*, and *C/c*. These numbers can be integers and fractions. For instance, the coordinates of the origin are 0 0 0, whereas the coordinates of the center point in the *BCC* lattice are 1/2 1/2 1/2. See also *lattice basis*.

lattice site See *lattice point*.

lattice void Position between *lattice sites*. It is also called *interstice*.

Laue diffraction pattern Pattern obtained by the *transmission* or *back-reflection Laue method*. It consists of regularly arranged *diffraction spots* corresponding to the reflections from various *lattice planes* of a *single crystal*. This pattern is used for determining the orientation of *single crystals* or coarse *grains*, as well as for *x-ray structure analysis*.

Laue equations Set of three equations describing directions of the x-ray beam diffracted from three non-coplanar atomic rows, i.e., on a three-dimensional lattice of diffraction centers.

Laue method X-ray technique using *white radiation* to determine orientation and *lattice* of *single crystals* wherein the diffraction pattern is registered on a flat film or screen.

Laves phase *Intermediate phase* of a varying *composition* close to AB_2 where both A and B are metals whose *atomic radii* relate as ~1.255. Laves phases are characterized by *coordination number* from 12 to 16.

ledeburite *Eutectic structure* in Fe–Fe_3C *system* occurring because of an *eutectic reaction* shown by the ECF line in Fe–Fe_3C diagram:

$$L_{4.3} \leftrightarrow \gamma_{2.08} + Fe_3C$$

The carbon content (in *wt%*) in the *L*- and *γ-phases* is shown by subscripts. It is named after German metallurgist A. Ledebur.

ledeburitic steel *Alloy steel* in which a small amount of *ledeburite* is present after *solidification*. The formation of ledeburite at carbon concentrations as low as ~1.5 wt% results from the influence of *alloying elements* on the carbon concentration range where the *eutectic reaction* takes place.

lenticular martensite See *acicular martensite*.

lever rule Rule used to find amounts of *equilibrium phases* in a two-*phase system*. The ends of the *tie line*, *ab*, in the $(\alpha + \beta)$ field of the *binary system* A–B show chemical *compositions* of α- and β-phases in a two-phase alloy X at temperature *T* (see Figure L.2). Let the tie line be a lever and the point *x*, denoting the composition of the alloy considered, its fulcrum. The lever will be in equilibrium if

$$M_\alpha/M_\beta = xb/ax$$

where M_α and M_β are the masses of α- and β-phases, respectively, assumed to concentrate at the ends of the lever. It follows from the rule that

$$M_\alpha/M_X = xb/ab$$

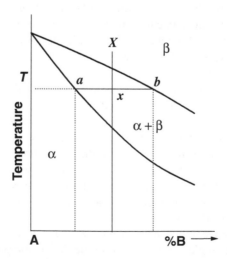

FIGURE L.2 Part of a binary phase diagram with the tie line *ab* in a two-phase field.

and

$$M_\beta/M_X = ax/ab$$

where M_X is the alloy mass, and all the compositions are expressed in wt%. The rule, in a slightly modified form, is applicable to *ternary systems,* provided an isothermal section of the corresponding *phase diagram* is considered. The rule is also used for determining the amounts of *microconstituents.* For example, the amount of *eutectoid* can be calculated as the amount of the high-temperature phase undergoing the *eutectoid decomposition.*

lineage structure See *cellular substructure.*

linear absorption coefficient Characteristic, μ, of the intensity reduction of radiation transmitting through a substance. It is defined by the equation:

$$I_t/I_0 = \exp(-\mu t)$$

where I_t and I_0 are the intensities of the incident and transmitted beams, respectively, and t is the transmitted thickness. It is supposed that there is no interaction between the incident and scattered radiation inside the sample (see *extinction*). See also *absorption coefficient.*

linear defect One-dimensional *crystal defect,* e.g., *dislocation* or *disclination.* The linear size of these defects in one dimension is considerably greater than the *atomic size,* and is commensurable with it in two other dimensions.

linear growth rate Rate of the *interface* migration at a constant temperature. Besides the *driving force,* the linear growth rate depends on the interface geometry, and because of this, can be strongly *anisotropic* (see, e.g., *diffusional transformation* and *Widmannstätten ferrite*). If the rate is *isotropic,* its magnitude can be estimated by $dD_{max}/2dt$, where D_{max} is the maximum linear size of *precipitates* or *grains,* and t is time. Linear growth rate can also be found during *in situ observations* with the aid of *PEEM* or a *hot-stage microscope.*

line broadening See *instrumental line broadening* and *intrinsic line broadening.*

liquation See *macrosegregation.*

liquid crystal Liquid organic compound, or a mixture of compounds, characterized by the *anisotropic* properties resulting from a certain preferred orientation of its non-*equiaxed* molecules. Anisotropic liquid, known also as *mesomorphic phase,* exists in a certain temperature range. Some liquid solutions of organic compounds can be also anisotropic. See *nematic, smectic,* and *cholesteric crystals.*

liquid-phase sintering *Sintering* of particulate compacts with *dopants* that melt during *firing.* The liquid phase accelerates the densification if it partially dissolves solid-phase particles, can wet the particles, and is characterized by an increased *diffusivity.* Since such a liquid penetrates between the particles, it facilitates densification by dissolving small particles (see *Thomson–Freundlich equation*) and rearranging the remaining ones. In cases in which all the *grains* are surrounded by the liquid phase, *grain coarsening* can be accelerated by preferentially dissolving the small grains, whereas the pore elimination may proceed rapidly enough due to a high diffusivity of the liquid phase. An increased diffusivity of liquid films between grains can trigger *abnormal grain growth* (see *solid-state sintering*).

liquidus Locus of *solidification temperatures* or that of chemical *compositions* of a liquid phase in equilibrium with solid phases in the *system* concerned. In *binary phase diagrams*, the locus is a line, and in *ternary* diagrams, it is a surface.

logarithmic creep Empirical time dependence of creep *strain*, ε, at temperatures $<0.3\ T_\mathrm{m}$ when *dynamic recovery* is negligible and different creep stages are indistinguishable:

$$\varepsilon = ac\ \ln(mt + 1)$$

where a, c, and m are constants, and t is time. Logarithmic creep is independent of temperature and stress magnitude.

Lomer–Cottrell barrier/lock *Sessile dislocation* in *FCC crystal lattice* formed through interaction of two leading *Shockley partials* that *glide* over intersecting *slip planes* toward one another. The resulting new *partial dislocation* lies along the intersection line; it also connects two *stacking fault* ribbons with two other Shockley partials remaining on their initial slip planes. Such a stair-rod dislocation cannot glide and so impedes the glide motion of the other dislocations over these planes.

long-range Relating to distances exceeding an *interatomic spacing* in the solid state.

long-range order Feature of *atomic structure* revealing itself in an arrangement of atoms (ions) over the *sites* of certain *crystal lattice*. This should not be confused with the *long-range ordering* in *solid solutions*.

long-range ordering See *order–disorder transformation*.

long-range order parameter In an *ordered solid solution*, a quantity defined as

$$S = (p - c)/(1 - c)$$

Here, p is a probability that an atom of one *component* occupies a position in its "own" *sublattice* (*antiphase domains* are not taken into account), and c is the atomic fraction of this component. The parameter S increases with the decreasing temperature, and at a perfect order $S = 1$. However, $S < 1$ at any temperature in ordered solutions whose *composition* is nonstoichiometric. See *order–disorder transformation*.

Lorentz factor Quantity characterizing the angular dependence of *integrated intensity* of the diffracted x-ray beam in the *powder* and the *rotating crystal methods*:

$$L = 0.25/(\sin^2\theta\ \cos\theta)$$

where θ is the *Bragg angle*. Lorentz factor is taken into account in *x-ray structure analysis*.

Lorentz microscopy *TEM* technique for studying *domain structures*. An interaction between the primary electrons and the magnetic field in *magnetic domains* makes the domains visible. The Lorentz microscopy can be used for *in situ* studies of the domain structure alterations on magnetization.

low-angle boundary *Grain boundary* with a *disorientation* angle smaller than ~15°. In *symmetric tilt* boundaries, *geometrically necessary dislocations* are *primary dislocations* of *edge* type, the same *sense*, and the same *Burgers vector*. They are arranged under one another and form a two-dimensional *dislocation network* called a *dislocation wall*. A *disorientation* angle, Θ, of such a boundary can be calculated as follows:

$$\Theta = b/h$$

where b is the Burgers vector, and h is the average distance between dislocations in the wall. In *asymmetric* tilt boundaries, along with the previously mentioned geometrically necessary dislocations, there are additional families of edge dislocations affecting the *grain boundary orientation*. A *twist* boundary is formed by a two-dimensional square net of *screw dislocations* having the same *sense* in each of the arrays of parallel dislocations. Its disorientation angle can be derived from the previous formula, where h is now the average distance between the dislocations in one of the arrays. Due to their elastic interactions, the dislocations in such a net usually rearrange into a two-dimensional hexagonal net. A stress field of low-angle boundaries extends at distances not greater than h. Low-angle boundary is also referred to as small-angle boundary or subboundary.

low-energy electron diffraction (LEED) Technique for studying a periodic *atomic structure* of clean surfaces and monoatomic *adsorption* layers by looking at *diffraction patterns* of low-energy electrons (1–5 keV) *elastically scattered* from the surface.

lower bainite *Microconstituent* evolving upon the transformation of *undercooled austenite* in a lower part of *bainitic range* and consisting of plate-like *ferrite* crystals and elongated *carbide* particles precipitating mostly inside the ferrite grains and partially between them. For the atomic mechanism of its formation, see *bainitic transformation*.

lower yield stress Magnitude of *tensile stress* corresponding to the *yield point elongation*. See *sharp yield point*.

low-temperature thermo-mechanical treatment See *ausforming*.

Lüders band *Macroscopic* area of a sample (article) where *plastic deformation* commences at a *stress* equal to the *upper yield stress* and completes at the *lower yield stress*, whereas the other areas remain elastically deformed. The Lüders band crosses the entire sample and thus, many *grains*. The occurrence of the bands and their spreading out is accompanied by an increase in macroscopic *strain* (see *yield–point elongation*). The spreading of Lüders bands is associated with the *glide* motion of the *dislocations* released from *Cottrell atmospheres* and *precipitate particles* or produced by *dislocation sources* or both. The surface of the sample (article) with the Lüders bands becomes roughened like *orange peel*.

Lüders strain See *yield point elongation*.

M

M_d temperature Temperature below which a *strain-induced martensite* occurs under the influence of *plastic deformation*; it is higher than the M_s^σ temperature.

M_s^σ temperature Temperature below which a *stress-assisted martensite* occurs; it is higher than M_s and lower than M_d temperature.

macrograph Photograph of *macrostructure*.

macroscopic stress *Residual stress* acting at distances significantly greater than the *mean grain size*.

macrosegregation Heterogeneity revealing itself in different chemical *composition* and, sometimes, in different *phase constituents* in various parts of an *ingot* or *casting*. Macrosegregation occurs on *solidification* and is always nonequilibrium. Its formation can be explained in the same way as that of *coring*, but in the case of macrosegregation, s_1, s_2, and s_3 (see Figure M.1) show the compositions of different ingot parts. Macrosegregation is also referred to as *major* or *zonal segregation* or as *liquation*.

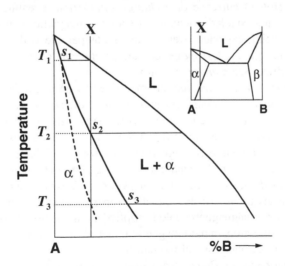

FIGURE M.1 Occurrence of macrosegration in alloy X in the course of solidification. Points s_1, s_2, and s_3 show compositions of the ingot parts solidified at temperatures T_1, T_2, and T_3, respectively. For further details, see *coring*, Figure C.3.

macrostructure *Structure* observed with the unaided eye or with a magnifying glass (i.e., at *magnifications* up to ~20×).

magnetic crystalline anisotropy Orientation dependence of the internal magnetic energy in *ferromagnetic* or *ferrimagnetic single crystals*. The *lattice direction*, along which the energy is the smallest, is called the *easy magnetization direction*. The angular dependence of the energy is described by a trigonometric series with coefficients known as constants of magnetic anisotropy. Magnetic materials with a low constant of magnetic anisotropy are soft magnetic, and those with a high constant are hard magnetic.

magnetic domain In *ferromagnetic* and *ferrimagnetic* materials, an area *spontaneously* magnetized up to saturation in the absence of an external magnetic field and at temperatures below T_C. The magnetization vector inside a domain lies close to the *easy magnetization direction*. In single-phase materials, the domain size is usually smaller than the *grain size*. However, in strongly *textured* materials, e.g., in $\underline{Fe}$–Ni alloys with a *cube texture*, or in $\underline{Fe}$–Si alloys with the *Goss texture*, the domain size can be greater than the grain size. In multiphase materials with small magnetic particles, the domain size depends on the particle size, and in some cases, the particles can be *single-domain*.

magnetic force microscope (MFM) Device analogous to *STM* and *AFM*. It measures magnetic stray fields on the surface of a magnetic specimen and allows an *in situ* investigation of the *domain structure* in the course of magnetization.

magnetic ordering Occurrence of *exchange interaction* revealing itself below T_C or T_N in a coupled orientation of atomic magnetic moments, the orientation being extended over distances comparable with the *grain* size. See *ferromagnetic, ferrimagnetic*, and *antiferromagnetic*.

magnetic structure See *domain structure*.

magnetic texture *Domain structure* characterized by an increased volume fraction of *magnetic domains* (in single-*phase* materials) or *single-domain particles* (in multiphase materials) with magnetization vectors parallel or antiparallel to a certain specimen direction. Because of this, a *polycrystal* with magnetic texture is magnetically *anisotropic*. Magnetic texture can be induced by *crystallographic texture* (e.g., by the *Goss texture* in Fe–Si *alloys* or *cube texture* in Fe–Ni alloys) or by *thermo-magnetic treatment*. Magnetic texture improves magnetic properties of soft magnetic materials when an external magnetic field is applied along the previously mentioned specimen direction. In hard magnetic materials, magnetic texture increases coercive force and residual induction.

magnetic transformation *Phase transition* wherein a *magnetically ordered phase* (e.g., a *ferromagnetic* one) transforms into a magnetically disordered, *paramagnetic* phase; or a magnetically ordered phase transforms into another phase with a different magnetic order (e.g., ferromagnetic phase ↔ *antiferromagnetic* phase). Magnetic transformation evolves as *second-order transition*.

magnification In *optical* and *electron microscopy*, the ratio of the linear size of a microstructural feature in the image to its true linear size, the magnification by N times being denoted by $N\times$.

major segregation See *macrosegregation*.

malleable [cast] iron Cast iron in which *temper carbon* occurs in the course of *graphitization annealing* of *white iron*. The metallic *matrix* in malleable irons can be ferritic or pearlitic, depending on the cooling rate from the annealing temperature as well as on the *alloy composition*.

maraging steel *Alloy steel* comprising 18–25 wt% Ni, <0.1 wt% C, and other *alloying elements* (e.g., Co, Mo, and Ti) with temperatures $M_d \cong 100°C$ and $M_s < 20°C$. After normalization, this steel is *austenitic*. *Plastic deformation* at ambient temperatures (i.e., below M_d) leads to the formation of a ductile *martensite*, its ductility being a result of low carbon content. Reheating the martensite below the A_s *temperature* (up to 400–500°C) results in the *precipitation* of *intermediate phases* accompanied by a significant strengthening (see *age hardening*). The degree of the strengthening depends on the *dislocation density* in the deformed martensite, as well as on the size and number of precipitates nucleating at the *dislocations* (see *heterogeneous nucleation*).

marquenching See *martempering*.

martempering *Heat treatment* consisting of cooling an article at two different rates: first, at a rate exceeding the *critical cooling rate*, from an *austenitic range* to a temperature slightly above M_s, and afterwards, when the temperature over the article's cross-section becomes uniform, at a much lower rate, to a temperature below M_f. Such a treatment diminishes *thermal stresses* induced by a rapid cooling and prevents thermal cracks. This treatment is also referred to as marquenching.

martensite *Metastable phase* occurring as a result of *martensitic transformation*. Since the transformation is *nondiffusional*, martensite (M) is of the same *composition* as the parent phase. M *crystallites* have a *coherent interface* with the parent phase, which reveals itself in a certain *orientation relationship* between the *lattices* of M and the parent phase. Since this interface is *glissile*, the crystallites can grow at a rate close to the speed of sound. They stop growing when they encounter the *grain boundaries* of the parent phase or other M crystallites. M crystallites are plate-like or lath-like. Their *substructure* is characterized by several *transformation twins* or by a high *dislocation density*, both of which are caused by an inhomogeneous deformation of M in the course of its *nucleation* and growth. M is named after German scientist A. Martens. See *steel martensite* and *titanium martensite*, as well as *lath martensite* and *acicular martensite*.

martensite finish temperature (M_f) Temperature at which the *martensite* formation, upon continuous cooling, is practically completed. This temperature depends on the parent phase *composition*.

martensite start temperature (M_s) Temperature at which *martensite* starts to appear upon cooling a parent phase. M_s depends strongly on the *composition* of the parent phase, as well as on the material prehistory (e.g., a decrease of *grain size* in a tetragonal ZrO_2 decreases M_s) and slightly, if

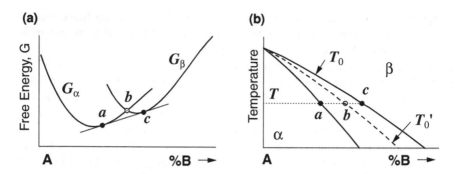

(a)

Free Energy, G

G_α G_β a b c

A   %B →

(b)

Temperature

T_0 β T a b c T_0' α

A   %B →

FIGURE M.2 (a) Free energies, G_α and G_β, of α and β solid solutions vs. concentration at temperature T. Intersection b corresponds to the equilibrium of metastable solutions of the identical composition. (b) Part of a binary phase diagram and corresponding equilibrium temperatures T_0 and T_0'. For further details, see *equilibrium temperature*, Figure E.1.

at all, on the cooling rate, provided the latter is higher than the *critical cooling rate*. In Ti *alloys* with β-*stabilizers*, M_s lies within the (α + β)-phase field in the corresponding *phase diagram*. M_s is always lower than the *equilibrium temperature*, T_0' (see Figure M.2), in many cases, by several hundreds of degrees.

martensitic range   Temperature range between M_s and M_f.

martensitic steel   *Alloy steel* consisting, after *normalizing*, predominately of *martensite*. This is due to the influence of *alloying elements* decreasing a *critical cooling rate* for *austenite* → martensite transformation.

martensitic transformation   *Phase transition* in the solid state with the following main features: it is a *shear-type transformation*, and the *shear strain*, necessary to transform the *lattice* of the parent phase into that of martensite, is ~0.2. The occurrence of *martensite* crystallites is accompanied by shape changes manifesting themselves in relief effects on the specimen surface, polished before the martensitic transformation (MT). In the course of MT, the atoms move cooperatively from their positions in the parent phase lattice into their positions in martensite. As a result, they retain the same neighbors as in the parent phase, and *lattice defects* (e.g., *dislocations*) are inherited by the martensite. The chemical *composition* of martensite is the same as that of the parent phase. MT commences below M_s (it is lower than the corresponding *equilibrium temperature*, T_0'; see Figure M.2), develops upon cooling, and completes below M_f. Since at a constant temperature between M_s and M_f it hardly evolves, MT is called *athermal* transformation. In some *systems*, however, an increase in the volume fraction of an *isothermal* martensite is observed, although a significant fraction of the parent phase remains untransformed. Either *plastic deformation* below the M_d temperature or *strain* below the M_s^σ temperature promotes MT, M_d and M_s^σ being higher than M_s (see *stress-assisted* and *strain-induced martensite*). MT in some alloys can be *reversible* (see *thermoelastic martensite* and *shape memory effect*). The *nucleation* of martensite is definitely *heterogeneous*, even though the majority of its nuclei normally escape detection.

mass absorption coefficient Characteristic of the *absorption* of radiation. It is equal to μ/ρ, where μ is a *linear absorption coefficient*, and ρ is the material density. See also *absorption coefficient*.

massive martensite See *lath martensite*.

massive transformation *Diffusional transformation* producing a new *metastable phase* of the same *composition* as the parent phase, i.e., evolving without *solute* partitioning. It is observed at an increased cooling rate, as well as at an increased *undercooling*. New grains *nucleate* primarily at the *grain boundaries* of the parent phase and grow at a relatively high rate because the diffusion paths are short (of the order of the *interface* thickness). The new grains are of an *equiaxed* shape and free from *lattice defects*. Massive transformation is, in many aspects, identical to *diffusional allotropic transformation*.

master alloy See *alloying composition*.

matrix In materials science, a *microconstituent* forming a continuous background as, e.g., a *plastically deformed* matrix in the case of *primary recrystallization*, an *austenitic* matrix in the case of *martensitic transformation*, a *solid solution* matrix in the case of *precipitation*, etc.

matrix band Part of a *plastically deformed grain*, adjacent to a *deformation band*. In contrast to the latter, the orientation across the matrix band does not change noticeably.

Matthiessen's rule Empirical rule according to which the contribution of both the *solutes* and *lattice defects* in electrical resistivity of *metallic* materials is temperature-independent. It is used for the estimation of the material's purity by measuring *residual electrical resistance*.

M-center *Color center* occurring as a result of electron trapping by an agglomerate of two *vacancies* in a cation *sublattice* and one vacancy in an anion sublattice.

mean [grain/particle] size Linear size averaged over the entire *grain* (particle) population. The quantity received by averaging over the number fractions can be quite different from that received by averaging over the volume fractions, especially in the case of *inhomogeneous microstructures* (e.g., at *duplex grain size*). Most techniques of *quantitative metallography* yield the volume-averaged mean size.

mechanical alloying Producing *alloy* powders without melting by a prolonged milling in high-energy ball mills. *Solid solutions* of insoluble *components* and strongly *supersaturated solid solutions,* as well as homogeneous mixtures of metallic and nonmetallic insoluble particles can be obtained by this method. The powder is used for producing *precipitation-strengthened* and *ODS* alloys.

mechanical anisotropy In materials science, anisotropy of *mechanical properties* resulting from *crystallographic texture*, or induced by *inhomogeneous microstructure* (e.g., by *carbide stringers* or *banded structure*), or both.

mechanical property Material response to any external influence causing *plastic deformation* or destruction of a sample, e.g., strength, toughness, ductility, etc. Almost all mechanical properties are *structure-sensitive*.

mechanical stabilization [of austenite] Decrease in the *martensite* amount after *quenching* plastically deformed samples in cases in which the parent *austenite* was subjected to *plastic deformation* above M_d. The effect can be explained by the influence of deformation on martensite *nucleation*.

median size Quantity equal to half the difference between maximum and minimum sizes. Since the *grain* (particle) *size distribution* is always skewed, the median grain (particle) size is greater than the *most probable size*.

melt spinning Procedure for producing thin ribbons by pouring a narrow melt jet onto a cylindrical surface of a rotating water-cooled wheel. In this method, the cooling rate can achieve ~10^6 K/s. Ribbons of *metallic glasses* can be obtained using this technique because the cooling rate is greater than *critical*. Melt spinning is also referred to as roller quenching.

melting point/temperature (T_m) Temperature when melting begins upon heating. It is shown by the *solidus* line (surface) in a *phase diagram*, and is usually expressed in K.

mesomorphic phase/mesophase *Stable* or *metastable phase* that exists in a certain temperature range and possesses the features characteristic of the phases existing just above and just below the range. For instance, *liquid crystals* are mesomorphic phases; they exist above the *melting point* and are liquid, yet simultaneously *anisotropic*, which is typical of *crystalline* phases.

metadynamic recrystallization Growth of *strain-free grains* in *hot-worked* material upon its subsequent *annealing*. After hot deformation, the nucleated *recrystallized* grains cannot grow because the cooling was rapid. Metadynamic recrystallization differs from *static recrystallization* in that there is no *incubation period*. Depending on the deformation conditions, metadynamic recrystallization can develop either in the whole deformed article or in a part of it.

metal ceramic Hard *composite* obtained by *sintering* compacts containing *carbide* particles, e.g., tungsten carbide, and *metallic* powder, e.g., nickel or cobalt.

metallic bond Bond exerted by free *valence* electrons shared by all the atomic nuclei. These electrons form an electron gas that holds the nuclei together and restrains their repulsion.

metallic crystal Crystal whose atoms are connected by a *metallic bond*; in transition metals, the interatomic bond can be partially *covalent*. Since the metallic bond is *isotropic*, *crystal lattices* of metals are characterized by high *atomic packing factors* (see *FCC*, *HCP*, and *BCC structures*).

metallic glass Metallic *alloy* with *amorphous atomic structure* containing, along with metals, certain nonmetallic *components*. It can be obtained by a rapid cooling from the liquid state, e.g., by *melt spinning*. Metallic glasses are characterized by an increased strength, high corrosion resistance, and excellent magnetic properties, as well as by the *isotropy* of these properties. See also *glassy phase*.

metallic radius *Atomic radius* in *metallic crystals* having *coordination number* 12 and formed by the atoms of one chemical element only. The radius is defined as half the minimum *interatomic spacing*, i.e., it is measured along the *close-packed lattice direction*. In crystals with another coordination number, the radius can be calculated using tabulated correction coefficients.

metallographic examination *Microstructure* study of polished or polished and etched *metallographic samples* using *optical microscopy, SEM,* or *PEEM*.

metallographic sample/section Plane sample for examining the *microstructure* of opaque materials by *optical microscopy, SEM,* or *PEEM*. The final operations of the sample preparation include mechanical, chemical, or electro-polishing and *chemical, electrolytic, thermal, color,* or *ion etching*. In some cases, microstructure can be observed on a polished surface using, e.g., *polarized-light microscopy*.

metastable β alloy See β *Ti alloy*.

metastable β-phase ($β_m$) In Ti alloys, β-*phase* occurring on *aging* from *titanium martensite*.

metastable phase Phase not associated with the absolute minimum of *free energy* of a thermodynamic *system*. Metastable phase occurs, provided the atomic mobility is restricted (as, e.g., *martensite, cementite*) and the thermodynamic barrier necessary for its *nucleation* is low in comparison to the *stable phase* (as, e.g., *cementite* in *cast irons*; see *graphitization*), or due to short diffusion paths necessary for its *nucleation* and growth (as, e.g., in *massive transformation*). A nontransformed phase below its equilibrium temperature can also be considered metastable if its transformation into more stable phases does not start yet, or if the commencement of the transformation is kinetically constrained.

metastable state Thermodynamic state of an isolated *system* characterized by excess *free energy* in respect to its *thermodynamic equilibrium*. There are three types of metastable state: a structural one associated with *metastable phases*, both *crystalline* and *amorphous*; a morphological one associated with various *crystal defects*, including *interfaces*; and a compositional one associated with a nonequilibrium *composition* of *stable phases*, as, e.g., *microscopic segregation*. A reduction of the first type of metastability gives rise to the occurrence (or disappearance) of certain *phase constituents* (see, e.g., *aging, tempering*). The metastability of the second and third types can be reduced without *phase transitions*, but only by *thermal activation* triggering the atomic displacements. For instance, *annealing* of *plastically deformed* materials results in *recovery, recrystallization,* and *grain growth* without any phase changes; all these processes reduce the excess free energy associated with *crystal defects*. Another instance of the reduction of morphological metastability is the *Ostwald ripening* leading to *coarsening* of *precipitated particles* and thus, to a reduction of the *interfacial energy* per unit volume not accompanied by any phase transition. The compositional metastability can be removed, or at least reduced, by *homogenizing*.

microalloying See *doping*.

microanalysis Chemical analysis of a small sample area.

microband See *deformation band*.

microconstituent In multiphase materials, a part of a *microstructure* consisting of either single-*phase grains* (e.g., *primary crystals*, *proeutectoid ferrite*) or crystallites of different phases arranged in a characteristic pattern (as, e.g., in *eutectoid colonies*). In single-phase materials, there can be more than one microconstituent if the microstructure is *inhomogeneous*.

microdiffraction Electron diffraction pattern from a small area (see *selected area diffraction pattern*).

micrograph Photograph of an image received by *optical* or *electron microscopy*.

microprobe X-ray or electron-beam microanalyzer used for chemical analysis.

microscopic stress *Residual stress* whose magnitude can be different in neighboring *grains*.

microsegregation Chemical heterogeneity inside *grains* observed in *solidified* articles. See *coring*.

microstrain Elastic deformation induced by microscopic stresses.

microstructure *Structure* characterized by the size, shape, volume fraction, and arrangement of *grains* of different *phases* or of a single phase. It is usually observed by *optical microscopy* and, in the case of exceptionally fine grains and *crystal defects*, by *electron microscopy*. The formation of microstructure (M) is determined, along with *phase transitions*, by the *kinetics* of the transformations or other processes reducing *free energy*. For instance, in the $\underline{Fe}$–0.8 wt% C *alloy*, a slow cooling from temperatures in an *austenitic range* results in the *pearlite* formation, whereas a rapid cooling leads to the *martensite* formation. In single-phase materials, M can be *homogeneous* or *inhomogeneous*, depending on the *grain size distribution* and on the spatial arrangement of grains of different sizes and shapes. In multiphase materials, M is described by the type and volume-fraction of *microconstituents*: e.g., *second-phase precipitates* and *matrix; proeutectoid ferrite* and *pearlite*; grains of different phases in *dual-phase* microstructure. M is always *metastable*, which is a result of either an increased *dislocation density* (as, e.g., in *plastically deformed* materials) or an increased *specific area* of *grain* and *phase boundaries*. Thus, under the influence of *thermal activation*, M can *spontaneously* change as, e.g., in *recrystallization*, *grain growth*, *spheroidization*, *particle coarsening*. See also *metastable state* and *substructure*.

microtexture Description of either all the possible *disorientations* between the grains in a *polycrystal* (see *misorientation distribution function*) or the disorientations between the neighboring grains only (see *grain-boundary character distribution*).

midrib See *acicular martensite*.

Miller indices Notation of *lattice planes* in a three-axis coordinate system whose axes coincide with the *crystal axes*. Miller indices can be found as follows. Take a lattice plane not passing through the origin and intersecting axes *x*, *y*, and *z* at distances *H*, *K*, and *L* from the origin; in cases in which the plane is parallel to a certain axis, the intercept length is assumed infinite. The numbers a/H, b/K, and c/L, where *a*, *b*, and *c* are *lattice constants*, can be fractions and/or integers. The Miller indices of the lattice plane are these numbers converted into a set of the lowest integers, *u*, *v*, and *w*, and the plane is denoted by the indices in parentheses: (*hkl*). For instance, if a plane is parallel to both the *x*- and *y*-axes and intersects +*z*-axis, its Miller indices are (001), and if it intersects the −*z*-axis, they are ($00\bar{1}$). There can be several lattice planes denoted by the same indices *h*, *k*, and *l*, arranged differently and having different signs (see *form*).

Miller–Bravais indices Notation of *lattice planes* in *hexagonal lattice* using a four-axis coordinate system. The necessity of the Miller–Bravais indices is derived from the fact that the *Miller indices* for identical planes in hexagonal lattice, except for a *basal plane*, are different: e.g., *prism planes* of the same order are denoted in the Miller notation variously, i.e., by (100), (010), ($1\bar{1}0$), etc. In the Miller–Bravais notation, the same planes have identical indices, except for their signs and the order of their arrangement: ($10\bar{1}0$), ($01\bar{1}0$), (0110), etc. In the Miller–Bravais system, the origin lies at the center of the basal plane; three axes, a_1, a_2, and a_3, lie in the plane at 120° to each other, and the fourth axis, *c*, is perpendicular to the basal plane (see Figure M.3). If a plane has Miller indices (*HKL*), its Miller–Bravais indices (*hkil*) are: $h = H$, $k = K$, $i = -(K + L)$, $l = L$. Sometimes, the third index in the Miller–Bravais notation is substituted by a point: e.g., ($11\bar{2}0$) plane can be denoted by (11.0). See *form* for the notation of the plane family.

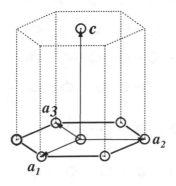

FIGURE M.3 Coordination axes in Miller–Bravais notation.

mirror plane One of the macroscopic *symmetry elements*. Mirror plane is a
   plane bringing a *point lattice* into self-coincidence with its mirror reflection in the plane.

miscibility gap Dome-shaped, two-phase field where two liquids, L_1 and L_2 (see
   binodal, Figure B.3) or two *isomorphous solid solutions*, α_1 and α_2, are
   present (see Figure M.4). The boundary of the field is known as *binodal*.
   The transformations $L \rightarrow L_1 + L_2$ or $\alpha \rightarrow \alpha_1 + \alpha_2$ can be considered as
   the occurrence of immiscibility in the liquid or solid state, respectively.
   Because of this, the field under the binodal is named a miscibility gap.

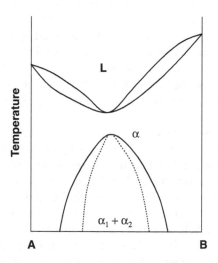

FIGURE M.4 Binary diagram with a miscibility gap in the solid state: α_1 and α_2 denote α solid
solutions of different compositions.

misfit dislocation In *partially coherent boundaries*, a dislocation that compen-
   sates *lattice misfit* (see Figure M.5). There are at least two systems of

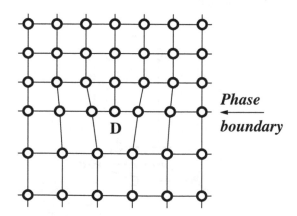

FIGURE M.5 Partially coherent phase boundary with a misfit dislocation D (scheme).

these dislocations at such a phase boundary. An interface containing misfit dislocations cannot *glide* because the *Burgers vectors* of these dislocations lie on the interface plane. If *epitaxy* is considered, misfit dislocations are referred to as *epitaxial*.

misfit parameter See *lattice misfit*.

misorientation See *disorientation*.

misorientation distribution function Description of all the possible *disorientations* between the grains in a *polycrystal* as a function of grain orientation in the Euler space (see *Euler angles*) or of their disorientation parameters in the Rodrigues space (see *Rodrigues vector*). It is also called microtexture.

mixed dislocation Dislocation whose *Burgers vector* is neither perpendicular nor parallel to its line. Mixed dislocation is usually considered consisting of *edge* and *screw* components.

mixed grain boundary *Grain boundary* whose *disorientation* axis is neither parallel nor perpendicular to its plane.

mode Value most frequently occurring in a *distribution curve*; it is also termed a most probable value.

modification In materials science, *grain refining* through the addition of certain substances to the melt prior to its pouring. Modification is often accompanied by changes in the *morphology* of *primary crystals* and *eutectics*. For instance, the modification of *Al*–Si *alloys* with sodium leads to a significant decrease in the size of Si crystals in the eutectic and to the disappearance of brittle primary crystals of silicon. The modification of *gray cast irons* with magnesium results in changes of the *graphite* shape from *flake*-like to *spheroidal*. The effect of modifying substances can be connected with an increase of the *nucleation rate* (see *inoculation*) or with their *adsorption* at the surface of primary crystals or both.

modulated structure In materials science, a *microstructure* observed after *spinodal decomposition*. It is characterized by an ordered arrangement of *coherent precipitates* along certain *lattice directions* in the *matrix grains*, the volume fraction of the precipitates being up to 50%.

modulus of elasticity See *Young's modulus* and *Hooke's law*.

modulus of rigidity See *shear modulus*.

moiré pattern Fringe *diffraction pattern* produced by electrons (x-rays) transmitting through two overlapping identical *crystal lattices*. The difference in the *lattice constants* (if the reflecting planes of the lattices are parallel) or the angle between these planes (if the lattices are rotated in respect to each other) can be calculated from the distance between the moiré fringes.

mol% Molar percentage used instead of *at%* when the *components* are compounds. It can also be denoted by mole% or $^m/_o$.

monochromatic radiation Any radiation with a wavelength varying in a narrow range.

monoclinic system *Crystal system* whose *unit cell* is characterized by the following *lattice parameters*: $a \neq b \neq c$, $\alpha = \gamma = 90° \neq \beta$.

monocrystalline Consisting of one *crystallite*.

monotectic reaction *Phase transition* following the reaction:

$$L_x \leftrightarrow \alpha_y + L_z$$

where α_y is the solid phase of *composition y*, and L_x and L_z are the liquid phases of compositions x and z, respectively (see Figure M.6); the right end of the arrow shows the reaction path upon cooling, and the left one shows the path up on heating. Independently of the reaction path, this reaction in *binary systems*, according to the *Gibbs phase rule*, is *invariant* and evolves at a constant temperature and pressure and constant x, y, and z.

monotectoid reaction *Phase transition* following the reaction:

$$\beta_x \leftrightarrow \alpha_y + \beta_z$$

where β_x, α_y, and β_z are the solid phases of *compositions x*, *y*, and *z*, respectively (see Figure M.6); the right end of the arrow shows the reaction path upon cooling, and the left one shows the path upon heating. Independently of the reaction path, this reaction in *binary systems*, according to the *Gibbs phase rule*, is *invariant* and evolves at a constant temperature and pressure and constant x, y, and z.

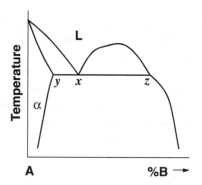

FIGURE M.6 Part of a binary phase diagram with monotectic reaction. In the case of mono-tectoid reaction, the phase fields on the diagram are of the same configuration, but the high-temperature phase is not L but a solid phase β.

mosaic structure In materials science, a conception used for the explanation of the *intrinsic broadening* of *x-ray diffraction lines*. It is based on the supposition that real crystals consist of slightly disoriented mosaic blocks having a perfect *lattice* and scattering x-rays independently. A decrease in the block size results in the line broadening. *Grains* in *nanocrystalline materials* and *subgrains* in materials with a greater grain size can be considered mosaic blocks. The nature of mosaic blocks in materials with increased *dislocation density* is uncertain.

most probable size See *mode*.

multiple cross-slip *Glide* motion of a *screw dislocation,* changing its *primary slip plane* many times by *cross-slip* over intersecting slip planes.

multiple jog See *jog.*

multiple slip Stage of *plastic deformation* in which several *slip systems* are active.

multiplicity factor Quantity affecting the diffracted beam intensity in the *powder method* or in *rotating crystal method.* Multiplicity factor is defined as the number of *lattice planes* of the same *form* having equal *interplanar spacings.* For instance, in a *cubic lattice,* it equals 6 for {100} planes because there are six equidistant {100} planes in the lattice. At the same time, in a *tetragonal lattice,* the multiplicity factor is 4 for (100) and (010) planes having the same interplanar spacings, and 2 for (001) planes having a different interplanar spacing.

N

Nabarro–Herring creep *Diffusional steady-state creep* at temperatures $>0.7\,T_m$ whose *strain rate*, $\dot{\varepsilon}$, is determined by the migration of *vacancies* through the grain body (and not over the *grain boundaries*, as in the case of the *Coble creep*):

$$\dot{\varepsilon} = a(\sigma/RT)D_v/\overline{D}^2$$

where σ is the *tensile stress*, D_v is the coefficient of *bulk self-diffusion*, $\overline{D}$ is the mean grain size, and R and T are the *gas constant* and the absolute temperature, respectively. The coefficient a depends on the grain shape:

$$a = 24A^{5/3}/(1 + 2A^2)$$

where A is the *grain aspect ratio* measured in the direction of the applied force.

NaCl structure *Crystal structure* wherein both the cation and anion *sublattices* are *FCC* and are shifted with respect to one another by $1/2a\langle100\rangle$, a being the *lattice constant* (Figure N.1). Another description of the structure may be the following: Cl^{1-} ions form an FCC sublattice, whereas the other sublattice is formed by Na^{1+} ions occupying all the *octahedral sites* of the first one.

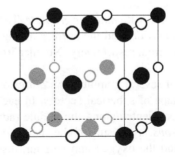

FIGURE N.1 Unit cell of NaCl structure; solid and open spheres show Cl^{1-} and Na^{1+} ions, respectively.

nanocrystalline Consisting of *grains* with a *mean size* smaller than ~100 nm.

natural aging *Decomposition* of a *supersaturated solid solution* or *aging treatment* at ambient temperatures.

N crystal See *nematic crystal*.

N* crystal See *cholesteric crystal*.

nearly special grain boundary *High-angle grain boundary* whose *disorientation* angle differs from that of a *special boundary* with approximately the same disorientation Θ by $\Delta\Theta \leq 15°\Sigma^{-1/2}$, where Σ is a *CSL* parameter defining the angle Θ at the special boundary.

Néel point/temperature (T_N, Θ_N) Temperature of the transformation *paramagnetic phase* $\leftrightarrow$ *antiferromagnetic* phase.

Néel wall In thin, magnetic films, a *domain wall* wherein magnetic moments lie parallel to the film surface, as distinguished from the *Bloch wall*.

nematic crystal *Liquid crystal* whose rod-shaped molecules fill the space densely and whose long axes are arranged approximately parallel to one another. These crystals are also called N crystals.

net plane See *lattice plane*.

Neumann band *Deformation twin* whose lenticular shape is distorted by *slip* inside it as well as in the adjacent *matrix*.

neutron diffraction Technique for studying *crystal structure* (especially that of substances containing *components* with a low *atomic mass*) and *crystallographic texture* of thick specimens, as well as *magnetic structure* using a monochromatic beam of thermal neutrons.

n-fold axis *Symmetry axis* with a rotation angle $360°/n$, where $n = 2, 3, 4$, or 6.

Nishiyama orientation relationship Orientation relationship between *martensite* (M) and *austenite* (A) in *steels* with an increased carbon content: $(101)_M \parallel (111)_A$ and $\langle 101 \rangle_M \parallel \langle 121 \rangle_A$. The *habit plane* in this case is $\{259\}_A$.

nitride *Intermediate phase* containing nitrogen as one of its major *components*. In systems with more than two components, *alloying elements* often dissolve in nitrides, forming complex nitrides or *carbonitrides*, if one of the components is carbon.

nodular [cast] iron *Cast iron* with *nodular graphite*. The *matrix* in nodular iron can be *ferritic* or *pearlitic*, and the corresponding iron names are ferritic or pearlitic nodular iron, respectively. Nodular iron is also referred to as ductile cast iron.

nodular graphite Joint of several graphite *crystallites* or branches of one crystallite having a shape of spherical sectors. In each of the sectors, the *c*-axis of its *lattice* is oriented along the nodule radius. This graphite morphology in *cast irons* is a result of the melt *modification* with definite substances that bind the oxygen and sulphur, dissolved in the melt and promoting the formation of the *flake graphite*. Nodular graphite is also referred to as spheroidal graphite.

nominal strain Magnitude of *tensile strain* calculated by taking into account the linear size of the nondeformed sample. It is also called engineering strain.

nominal stress Magnitude of *tensile stress* calculated as the external force divided by the initial cross-sectional area of the sample. It is also called engineering stress.

nondiffusional transformation Transition of a parent phase into a new one, often *metastable*, wherein atomic displacements are smaller than the *inter-atomic spacing* and proceed cooperatively, as in the course of *shear*, without *diffusion*. As a result, the same neighbors, as in the parent phase, surround the atoms in the *crystal structure* of the new phase, and the latter inherits all the *crystal defects* from the former. *Interfaces* between the crystallites of the parent phase and the new one are *coherent*, and the growth evolves as long as the interface retains its structure. The growing crystallites assume a plate- or lath-like shape, with their flat parts parallel to the *habit plane*. See also *shear-type* and *martensitic transformations*.

non-oriented [Material] characterized by a weak *crystallographic texture* or by a random *grain* orientation.

normal anisotropy Difference between the plastic properties over the sheet surface and along the normal to it. Normal anisotropy is quantitatively characterized by $\bar{r}$ -*value*.

normal grain growth *Grain* growth during which the *mean grain size*, $\bar{D}$, increases continuously due to the growth of the majority of grains. Normal grain growth (NG) proceeds by migration of *grain boundaries* to their centers of curvature under the influence of *capillary driving force* (see *curvature-driven grain growth*). Since the magnitude of $d\bar{D}/dt$ in the course of NG in single-*phase* materials is proportional to $1/\bar{D}$, the mean grain size increases during NG according to the equation:

$$\bar{D}^2 - \bar{D}_0^2 = k(t - t_0)$$

where the subscript 0 relates to the initial t, and $\bar{D}$, t is the annealing time, and constant k contains the *grain-boundary mobility* and *energy* averaged over all the boundaries. If $\bar{D} \gg \bar{D}_0$, the equation simplifies to

$$\bar{D} = kt^n$$

with $n = 1/2$. The n magnitude observed experimentally is usually smaller than 1/2, which is commonly (but not always) connected with the NG inhibition by certain *drag forces*. In the case $n < 1/2$, the constant k in the second equation is no longer a measure of the grain-boundary properties. In *crystalline ceramics*, NG can be controlled by the *diffusivity* of the intergranular liquid layers, by the reaction rate at the interface liquid–crystal, or by segregated impurity atoms (see *impurity drag*), as a result of which, n usually is ~1/3. During unimpeded NG, *microstructure* is *homogeneous* and the grain *size distribution* remains *self-similar*. If NG is inhibited, the shape of the grain size distribution changes significantly. In materials with a *duplex microstructure*, NG develops slower and the exponent in the previous equation is ~1/4. NG is also known as continuous grain growth.

normalizing *Heat treatment* of *steels*, including *austenitization*, followed by cooling in still air. It is also called normalization.

normal stress Stress constituent directed perpendicular to the plane section of a sample it acts upon.

nucleation First stage of any *first-order phase transition*. It consists in *spontaneous* occurrence of *crystallites* of a new phase. These crystallites (named *nuclei*) have the *atomic structure* and *composition* of the new phase and are separated from the parent phase by a *phase boundary*. Their occurrence is connected with overcoming a certain thermodynamic barrier. This explains why nuclei occur at some *undercooling* necessary to compensate the energy of the new *interface*, as well as the *elastic strain energy* associated with a *specific volume* difference between the nucleus and the parent phase (see *critical nucleus*). The tendency to decrease the energy of the new interface reveals itself in the formation of *coherent* or *partially coherent* phase boundaries. As a result, a definite *orientation relationship* is maintained between the new and the parent phases. In the solid state nuclei usually occur at *lattice defects* in the parent phase (see *heterogeneous nucleation*).

nucleation agent A substance intentionally added into the parent liquid with the aim of promoting *devitrification* (see *glass-ceramic*) on subsequent *annealing* the *glassy phase* obtained from the liquid. On heating the glassy phase, this substance provides homogeneously distributed small *crystals* facilitating crystallization, because the energy of their interface with the *nuclei* of a new crystalline phase is reduced. See *heterogeneous nucleation*.

nucleation rate Number of *nuclei* occurring in a unit volume of an untransformed parent *phase* per unit time.

nucleus In materials science, a small *crystallite* of a new *phase* with a *lattice* and, in many cases, *composition*, different than those of the parent phase. Nucleus (N) is separated from the parent phase by a *phase boundary*, and its size should be greater than critical (see *critical nucleus*). Its growth into the parent phase continuously reduces the *free energy* of the *system* concerned. The N shape and orientation with respect to the matrix *lattice* are affected by the distortions associated with a *specific volume* difference of the N and the parent phase, as well as by the *interfacial energy*. If N has *incoherent interfaces* and the *elastic deformations* due to lattice distortions are accommodated predominately in the matrix, the energy of the distortions is minimum, provided the N shape is plate-like. If N has *coherent* or *partially coherent interfaces*, its shape is governed by the compromise between the interfacial energy and the *elastic strain energy*. N is *equiaxed* in the case of small distortion energy, and plate-like in the opposite case.

numerical aperture In *optical microscopes*, a lens characteristic responsible for the *resolution limit* of the lens:

$$A_N = n \sin \alpha$$

where n is the refraction index of the medium between the object and the lens, and α is half the opening angle of the lens.

O

ω-phase *Metastable phase* with a *hexagonal crystal lattice* (c/a = 0.613) observed in α + β and β Ti *alloys* after *quenching* (known as athermal ω-phase) or, more commonly, after *aging treatment* below ~500°C. The phase occurs as a result of *shear-type transformation* β → ω The *orientation relationship* between ω- and β-phases is: $\{0001\}_\omega \| \{111\}_\beta$ and $\langle 2\bar{1}10 \rangle_\omega \| \langle 110 \rangle_\beta$. Small ω-phase precipitates (of *mean size* ~3 nm) have *coherent* interfaces, are *equiaxed*, and are arranged homogeneously.

octahedral interstice See *octahedral void.*

octahedral plane {111} plane in *cubic lattices.*

octahedral site See *octahedral void.*

octahedral void In *crystal structures,* a *lattice void* surrounded by six atoms forming the vertices of an octahedron. If the rigid spheres of equal radii represent the atoms, the radius of the void is the maximum radius of a nondistorted sphere arranging inside the void and touching the nearest atoms. In *FCC* and *HCP lattices,* the number of octahedral voids is equal to the number of atoms. Octahedral void is also referred to as octahedral site or octahedral interstice.

one-way shape memory effect Restoration of the initial shape of an object revealing *pseudoplasticity* upon cooling below M_f temperature. If the object is reheated through the A_s–A_f range, a reverse transformation of *martensite* into a parent phase takes place, which completely restores the original shape of the object. See also *reversibility.*

optical microscope Microscope used for *microstructure* study on flat opaque specimens in the visible light at *magnifications* from ~30 to ~1300×, with a *resolution limit* ~0.2 μm, and a *depth of focus* ~0.1 μm (at magnification ~1000×). In contrast to biological microscopes, the structure is observed in reflected light. A *bright-field* and *dark-field illumination*, oblique illumination, and illumination by polarized light (see *polarized-light microscopy* and *Kerr microscopy*), as well as *phase contrast,* can be used in optical microscopes. The microstructure image can be improved using an interference contrast. The image can be photographed or videorecorded.

orange peel In material science, surface roughening in drawn or stretched metallic sheets resulting from the *sharp yield point* phenomenon or the

Portevin–Le Chatelier effect. It can also be caused by a *coarse-grained structure* because *slip traces* in coarse grains are visible to the unaided eye. See also *Lüders band* and *stretcher-strain marking*.

order–disorder transformation/transition *Phase transition* in *substitutional solid solutions*. It consists in transforming the arrangement of *solute* atoms over the *lattice sites* from random (above the transition temperature) into regular (below the transition temperature) upon cooling or vice versa upon heating. In an ordered phase, the atoms of different *components* are preferably located at the sites of interpenetrating *sublattices* (see Figure O.1). Order–disorder transformation can proceed as a *first-order* or *second-order* transition, which can be traced by the temperature dependence of the *long-range order parameter*. See also *partially ordered solid solution*.

ordered solid solution *Substitutional solid solution* characterized by a regular arrangement of the *solute* and *solvent* atoms over different *sublattices* (see Figure O.1). The degree of the regularity is characterized by a *long-range order parameter*. See also *order–disorder transformation*, *partially ordered*, and *random solid solutions*.

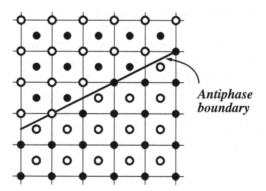

FIGURE O.1 Antiphase domains inside a grain of an ordered solid solution. Open and solid circles represent atoms of different components. They form sublattices that shift with respect to each other by neighboring domains.

orientation distribution function (ODF) Graphic description of *crystallographic texture* by grain orientation density in a three-dimensional space of *Euler angles* φ_1, Φ, and φ_2. Texture identification by a series of two-dimensional ODF sections, at different values of φ_2, is often much more precise and reliable than that by *pole figures*. ODF is obtained from the experimental data used for the pole-figure derivation. *Anisotropy of structure-insensitive properties* can be calculated using ODF.

orientation imaging microscopy (OIM) *SEM* technique using *EBSP* for determining and displaying orientations of different *grains* in *polycrystals*.

orientation relationship Relative orientation of two different *crystal lattices*; it is described by the indices of both the parallel *lattice planes* and the parallel *lattice directions*. If the indices are low, this can be considered as a proof of the presence of *coherent* or *partially coherent phase boundary*, at least at the *nucleation* stage.

orientation sphere See *stereographic projection*.

orientation spread See *texture scatter*.

Orowan loop *Dislocation loop* that surrounds a *precipitated* particle (or a *dispersoid*) in plastically deformed materials. It occurs when a gliding dislocation cannot shear a particle and is forced to bend over it. There can be several Orowan loops around each particle, and these loops contribute to *precipitation hardening* (or *dispersion strengthening*) because they repel the dislocations that follow those forming the loops.

Orowan mechanism Explanation of *precipitation hardening* and *dispersion strengthening* by bowing of *dislocations* during their *glide* motion through the gaps between *dispersed particles*. The bowing increases the *flow stress* by:

$$\Delta\sigma = \alpha Gb/\Lambda$$

where α is a coefficient, G is the *shear modulus*, b is the *Burgers vector*, and Λ is the mean spacing between the particles. If the particles are *coherent*, strengthening is determined by *particle shearing* at Λ smaller than a certain critical spacing.

orthoferrite. *Ferrimagnetic* oxide of *stoichiometry* $MFeO_3$ (M is a rare-earth element, Y, Ca, Sr, or Ba) and *perovskite crystal structure*.

orthorhombic system. Crystal system whose unit cell is characterized by the following *lattice parameters*: $a \neq b \neq c$, $\alpha = \beta = \gamma = 90°$.

Ostwald ripening. Increase of the *mean size* of *precipitates* with time at an increased temperature proceeding *spontaneously* in *saturated solid solutions* due to different solubility in the solution that surrounds precipitates of different radii (see *Gibbs–Thomson equation*). A diffusion flux, equalizing the *solute concentration*, results in both undersaturation of the solution surrounding smaller particles and supersaturation of the solution surrounding greater ones. The tendency to restore the equilibrium concentration results in dissolution of the smaller particles and growth of the larger ones. All this eventually reveals itself in an increase of the *mean particle size*.

overaging. Stage of *aging* characterized by progressive *coarsening* of the *precipitates* occurring at the first aging stages. Overaging decreases the effect of *precipitation hardening*.

oxide dispersion strengthened (ODS). Hardened by homogeneously distributed, *thermally stable* oxide particles (see *dispersion strengthening*). ODS materials are obtained mostly from *mechanically alloyed powders* (with the oxide particle size ~10 nm and oxide volume fraction from 3 to 10%), canned in a metallic container and consolidated by a multipass *warm deformation*. After the deformation, they possess an *ultra-fine grained* microstructure (with a mean grain size <0.1 μm) and are devoid of porosity. ODS alloys are characterized by a higher *creep* resistance than *precipitation-strengthened* alloys because oxides in the former are thermally stable, prevent the rearrangement of *dislocations* at increased temperatures, and thereby retard the *dislocation creep*. The oxides also retard the *subgrain* formation, subgrain growth, and *primary recrystallization*. If *oxides* form stringers after consolidation, a *coarse-grained microstructure* can be obtained by means of high-temperature *annealing*, the coarse grains being elongated along the stringers.

oxynitride *Solid solution* or an *intermediate phase* in a *system* whose *components* are nitrides and oxides. Typical examples of oxynitrides are *sialons*.

P

packet martensite See *lath martensite.*

packing factor See *atomic packing factor.*

paramagnetic Material having no magnetic moment of its own and slightly magnetized by an external magnetic field in the direction of the field.

partial dislocation Dislocation whose *Burgers vector* is smaller than the *translation vector* in the *crystal structure* concerned. This is the reason why the *glide* motion of partial dislocations distorts crystal structure and produces *stacking faults.* Partial dislocation is also called imperfect.

partially coherent interface *Phase boundary* where the *lattice misfit* cannot be accommodated by *coherency strains* (see *misfit dislocation*, Figure M.5). This, in turn, leads to the occurrence of *misfit dislocations* at the boundary. They weaken the strain field at the boundary, but the energy of partially coherent interface is higher than the energy of the *coherent interface.* Partially coherent interface is also termed semi-coherent.

partially coherent precipitate *Second-phase* particle whose *interface* with the *matrix* phase is *partially coherent*; at some areas, it can be *coherent.* A certain *orientation relationship* is observed between the *lattices* of the precipitates and the matrix phase.

partially ordered solid solution *Ordered solid solution* in which some atoms occupy sites in a "wrong" *sublattice*, and thus, a *long-range order parameter S* < 1. Such a *crystal structure* is observed, e.g., in cases in which the composition of the solution deviates from *stoichiometric* or at temperatures close to the temperature of the *order–disorder transformation.*

particle coarsening Increase in the *mean size* of *precipitates* (see *Ostwald ripening*) or the lamellae thickness in *colonies* (see *discontinuous coarsening*).

particle drag *Drag force* exerted by *dispersed particles* or *pores.* If the particles of the volume fraction f and the mean diameter d are arranged at the *grain boundaries* as, e.g., pores in *crystalline ceramics*, the drag force is:

$$\Delta g = a\gamma_{gb} f^{1/3}/d$$

where a is a coefficient, and γ_{gb} is the *grain-boundary energy* (it is supposed equal to the energy of the *phase boundary* between the particles and matrix). Pores inhibit grain boundary migration if they are relatively large and immobile; small pores can migrate together with the boundary and affect its motion to a lesser extent than immobile pores. In the case of uniformly distributed particles, the drag force is referred to as *Zener drag* and is proportional to *f/d*.

particle shearing One of the mechanisms of *precipitation hardening* by *coherent* or *partially coherent particles* that are cut by gliding *dislocations*. Particle shearing increases the *flow stress*, σ, by

$$\Delta\sigma = a(fd)^{1/2}$$

where *f* and *d* are the volume fraction and the mean diameter of the particles, respectively, and *a* is a coefficient. Particle shearing operates at an interparticle spacing smaller than critical. At a greater spacing, precipitation hardening is determined by the *Orowan mechanism*.

particle-stimulated nucleation (PSN) Occurrence of *recrystallization nuclei* at coarse (~1 μm) hard particles in heavily deformed materials. This is caused by a *lattice* rotation up to 20–30° in the deformed *matrix* in the vicinity of the particles, the rotation resulting from accumulation of *Orowan loops* at the particles in the course of *slip* deformation.

pearlite *Microconstituent* occurring in a *pearlitic range* because of an *eutectoid decomposition* in Fe–Fe$_3$C *alloys* on cooling from an *austenitic range*. Pearlite consists of *ferrite* and *cementite* (their weight proportion is 7.3:1, according to the *lever rule*), forming *pearlitic colonies*. Pearlite with a decreased *interlamellar spacing* is known as fine pearlite, or sorbite, and that with an increased spacing as coarse pearlite.

pearlitic cast iron *Gray iron* whose *microstructure* consists of *flake graphite* and a pearlitic *matrix*.

pearlitic colony/nodule *Equiaxed* complex consisting of two interpenetrating *single crystals* of *ferrite* and *cementite*, on *metallographic sections*, looking like alternate lamellae perpendicular to the colony interface. The colony grows via carbon diffusion, either through the *austenite* ahead of it or over its interface (see *coupled growth*). The *interlamellar spacing* in colonies, as well as their average size, decreases with a temperature reduction. Pearlitic colonies nucleate at *austenite grain boundaries,* as well as at interfaces between austenite and *proeutectoid ferrite* (in *hypoeutectoid steels*), or between austenite and *proeutectoid cementite* (in *hypereutectoid steels*). Since the growth rate of pearlitic colonies is *isotropic*, they are *equiaxed*.

pearlitic range Temperature range wherein *pearlitic transformation* evolves upon cooling from an *austenitic range*. The high-temperature limit of the range is A_1 temperature (see *baintic range*, Figure B.1). In *alloy steels*, the high-temperature limit can be shifted to higher or lower temperatures in comparison to *plain carbon steels*, depending on the effect of *alloying elements* on the austenite *thermodynamic stability* (see *ferrite-stabilizer* and *austenite-stabilizer*). Since the transformation is *diffusion-controlled* and evolves when both Fe and C atoms can move under the influence of *thermal activation*, the low-temperature limit of pearlitic range is a temperature below which the diffusion rate of Fe atoms is quite low. This limit in alloy steels is usually higher than in plain carbon steels, since the diffusion rate of many alloying elements is lower than that of iron.

pearlitic steel *Alloy steel* whose *microstructure* after *normalizing* is almost completely *pearlitic*. The microstructure can look like *pearlite* in some range of carbon *concentrations*, which can be explained by the fact that crystallites of *proeutectoid ferrite* or *proeutectoid cementite* are rather small and cannot be resolved on the pearlitic background.

pearlitic transformation *Eutectoid reaction* in *Fe–Fe$_3$C alloys* evolving in a *pearlitic range*:

$$\gamma_{0.8} \leftrightarrow \alpha_{0.025} + Fe_3C$$

The carbon content (in *wt%*) in the γ- and α-*phases* is shown by subscripts. The product of the transformation is *pearlite* or *sorbite*. Pearlitic transformation is *diffusion-controlled*, and its *kinetics* can be described by the *Johnson–Mehl–Kolmogorov equation*, provided *pearlitic colonies* are considered as *nuclei*. Pearlitic transformation in *alloy steels* is retarded, because many *substitutional alloying elements* decrease the carbon *diffusivity*. In addition, since the atoms of the alloying elements should redistribute between *ferrite* and *cementite*, their low diffusion rate increases the *incubation period* and decreases the rate of pearlitic transformation.

Peierls stress/barrier *Shear stress* per unit-length of a *dislocation*, necessary for its *glide* through a perfect *lattice* over a certain *lattice plane* and in a particular *lattice direction*:

$$\tau_p \propto [2G/(1 - \nu)] \exp(-2\pi w/b)$$

where *w* is the *dislocation width*, *b* is the *Burgers vector*, ν is the *Poisson ratio*, and *G* is the *shear modulus*. Peierls stress increases with the decreasing dislocation width that, in turn, decreases with an increase in the directionality and energy of interatomic bonds. Because of this, an increase in τ_p is observed in the following order: *FCC* or *HCP* metals → *BCC* transition metals → *ionic* or *covalent crystals*. This explains why the intrinsic brittlenes of ionic and covalent crystals is much greater than in *metallic* crystals. The magnitude of τ_p decreases with a decreasing *b* and increasing *interplanar spacing*, which explains why *densely packed planes* and *close-packed directions* form the commonly observed *slip systems*.

pencil glide *Slip* in *BCC* metals that proceeds over different *lattice planes* of ⟨111⟩ zone, ⟨111⟩ being the *slip direction*. This is a result of high *stacking-fault energy* in these metals, which makes the *cross-slip* quite pronounced.

perfect dislocation *Dislocation* with the *Burgers vector* identical to the *translation vector* in the *crystal structure* concerned. Thus, the *glide* motion of perfect dislocation does not distort the crystal structure, in contrast to *partial dislocations*.

peritectic reaction *Phase transition* that in a *binary system* (see Figure P.1) follows the reaction:

$$L_x + \alpha_y \leftrightarrow \beta_z$$

where L_x is the liquid phase of *composition x*, and α_y and β_z, are the solid phases of compositions y and z, respectively; the right end of the arrow shows the reaction path upon cooling, and the left arrow shows the path upon heating. In a *ternary* system, the reaction proceeds as follows:

$$L_a + \alpha_b \leftrightarrow \beta_c + \sigma_d$$

where L is a liquid, and α, β, and σ are solid phases, and the indexes relate to their compositions, or as:

$$L + \delta \leftrightarrow \omega$$

where L is liquid, and δ and ω are solid phases. According to the *Gibbs phase rule*, the first two reactions are *invariant* and thus, temperature, pressure, and phase compositions in the course of the reactions remain constant, whereas the phase compositions in the third one vary in the course of the reaction.

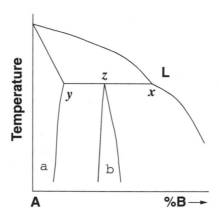

FIGURE P.1 Part of a binary phase diagram with a peritectic reaction. In the case of a peritectoid reaction, the phase fields are of the same configuration, but instead of a liquid phase, L, there should be some solid phase, γ.

peritectic temperature Temperature of *peritectic reaction* in the corresponding
phase diagram.

peritectoid reaction *Phase transition* that follows the reaction:

$$\gamma_x + \alpha_y \leftrightarrow \beta_z$$

where γ_x, α_y, and β_z are solid phases of *compositions x, y*, and *z*, respectively
(see Figure P.1); the right end of the arrow shows the reaction path upon
cooling, and the left one shows the path upon heating. Independent of the
reaction path, this reaction in *binary systems*, according to the *Gibbs phase
rule*, is *invariant* and evolves at a constant temperature and pressure and
constant *x, y,* and *z.*

perovskite Mineral of *composition* CaTiO$_3$.

perovskite [structure] type *Cubic crystal structure* of *ionic crystals* of *composition* ABO$_3$; it is identical to that of *perovskite.* In the structure, large A-
cations occupy the cube corners; O-anions occupy the cube face centers;
and there is a small B-cation at the center of the cube body. The following
cation charges usually correspond to the *stoichiometry* ABO$_3$: A^{2+} and B^{4+},
or A^{3+} and B^{3+}, although there may be other combinations. Many perovs-
kites undergo a *displacive polymorphic transformation*, e.g., a cubic form
of BaTiO$_3$ transforms into a *tetragonal* one in which the Ti^{4+} ion is slightly
displaced along $\langle 001 \rangle$ from its centered position. Thus, the tetragonal
structure is no longer centrosymmetric, which leads to the appearance of
a permanent electrical *dipole.* This transformation is also observed in
piezoelectric PbBaO$_3$, in *ferroelectric* Pb(Zr,Ti)O$_3$ and (Pb,La)(Zr,Ti)O$_3$,
known as PZT and PLZT, respectively, etc. Many oxides with perovskite
structure are semiconductors, and some others are *ferrimagnetics* (see
orthoferrite).

phase In thermodynamics and materials science, a part of a *system* consisting,
in *polycrystalline* materials, of many *crystallites* with the same *composition* and properties, as well as the same *crystal structure.* In multiphase
systems, different phases are in contact at *phase boundaries*, and the
crystal structure, composition, and properties change abruptly at the
boundaries. In contrast, the boundaries between *grains* of the same phase
(i.e., *grain boundaries*) are not taken into account. Sometimes, chemical
composition of some phases can be inhomogeneous, even inside one *grain*
(see, e.g., *coring*).

phase boundary Interface between the *grains* of different *phases.* In *crystalline
solids*, there can be *coherent, partially coherent,* and *incoherent interfaces.*

phase composition Complex of *phases* present in a *system.* The phases can be
both *equilibrium* (stable) or *metastable.* If they are stable, phase composition of an *alloy* can be found from the corresponding *phase diagram.*

phase constituent A given *phase* in a *heterophase system.*

phase contrast In *optical microscopy*, a technique for studying *microstructure*
on polished samples. Phase contrast results from the phase difference of
light beams reflected from areas of various chemical compositions or from

bulges and cavities of the surface pattern. Phase contrast in *electron microscopy* is often called diffraction contrast.

phase diagram Graphic representation of *thermodynamic equilibrium* in an *alloy system* consisting of one (single-phase system) or many *phases* (multiphase system). *Phase constituents*, their chemical *compositions,* and weight fractions at different temperatures and pressures, as well as the *phase transitions* accompanying temperature and pressure alterations, can be predicted using phase diagrams. The *kinetics* of phase transitions are not taken into account in phase diagrams. Phase diagram is also called constitutional or equilibrium diagram. See *tie line* and *lever rule.*

phase rule. See *Gibbs' phase rule.*

phase transition/transformation. Change in the number and nature of *phases* caused by a *free energy* reduction accompanying temperature or pressure alterations in a *system.* Phase transitions can evolve as *first-order* when *nuclei* of a new phase occur inside the parent phase and both of the phases coexist, or as *second-order* when the parent phase entirely transforms into a new one and there is no coexistence of the phases. Phase transitions can evolve as *diffusional* (reconstructive) or *nondiffusional* (*shear*, displacive, or *martensitic*), depending on the mechanism of the *phase boundary* migration. If the boundary migrates via the cooperative displacement of atoms smaller than an *interatomic spacing*, the transition is referred to as nondiffusional. If atoms pass the boundary individually and the magnitude of their displacements is not smaller than the interatomic spacing, the transition is referred to as diffusional.

photo-electron emission microscope (PEEM). Device for studying chemistry, *grain* orientation, and *microstructure* of flat, massive specimens using electrons emitted from the specimen's surface under the influence of a focused beam of high-intensity ultraviolet light (penetration depth ~10 nm). The lateral *resolution* of PEEM is ~15 nm. Heating attachment to PEEM makes it suitable for *in situ observations.*

physical adsorption. Bonding of *adsorbate* atoms and molecules to the *adsorbent* surface by the *van der Waals* or another weak bond, in contrast to *chemisorption.* It is sometimes called physisorption.

physical property. Material response to any external influence, except a chemical one or that causing *plastic deformation* or destruction of a sample.

physisorption. See *physical adsorption.*

piezoelectric. Material whose *lattice constant*, and thus the linear size of the body, changes under the influence of an external electric field and whose electrical polarization can be caused by application of an external load.

pile-up. Array of parallel *edge dislocations* of the same *sense* accumulated on a *slip plane* at an obstacle, such as a *precipitate, grain boundary,* or *interface.* The *stress* at the pile-up tip is $N\sigma$, where N is the number of dislocations in the pile-up, and σ is the *tensile stress.* Since N increases with the distance, L, between the obstacle and the *dislocation source* producing the pile-up, and because $L \leq \overline{D}/2$ ($\overline{D}$ is the *mean grain size*), an increase in $\bar{D}$ increases $N\sigma$ and thus decreases the *flow stress*. See *Hall–Petch equation* and *grain-boundary strengthening*.

pinning force See *drag force*.

pipe diffusion Diffusion along *dislocation cores*. Due to low *activation energy*, it is the main diffusion mechanism of both *host* atoms and *substitutional solutes* at temperatures lower than ~0.3 T_m.

plain carbon steel Steel containing no *alloying elements*.

planar anisotropy In materials science, anisotropy of *mechanical properties* in the sheet plane. The degree of planar anisotropy is quantitatively characterized by Δr-*value*.

planar defect In materials science, any defect of *crystal structure* (e.g., *free surface, phase boundary, subboundary, grain boundary, twin boundary, antiphase boundary, stacking fault*, etc.) whose linear size in two dimensions exceeds greatly the *atomic size* and is commensurable with it in the third dimension.

plastic deformation Residual deformation retained after removal of an external force. The same name is also applied to any processes producing residual deformation. See *deformation mechanism*.

plate martensite See *acicular martensite*.

point defect *Lattice* imperfection whose linear dimensions are commensurable with *atomic size*. Point defects comprise *vacancies, self-interstitials, foreign atoms, structural vacancies, antistructural atoms, antisite defects*, etc., as well as their associations (e.g., *divacancy, Schottky pair, Frenkel pair*, etc.). All the point defects distort the crystal lattice, which affects the evolution of *plastic deformation* (see, e.g., *solid-solution strengthening* and *diffusional plasticity*) and *diffusion* processes. These defects also provide distortions in the *band structure* (see, e.g., *acceptor, donor*, and *shallow impurity*).

point group See *symmetry class*.

point lattice See *crystal lattice*.

Poisson's ratio Negative ratio of the lateral and longitudinal *strains* in tensile (compression) samples subjected to elastic deformation.

polar net Projection of longitude and latitude lines of a *reference sphere* on its equatorial plane (see *stereographic projection*). It is also called equatorial net.

polarization factor Quantity characterizing the angular dependence of the intensity of the diffracted x-ray beam (in comparison to the intensity of the nonpolarized primary beam):

$$p = (1 + \cos^2 2\theta)/2$$

where θ is the *Bragg angle*. Polarization factor is taken into account in *x-ray structure analysis*.

polarized-light microscopy *Optical microscopy* using polarized light for increasing the image contrast. In this technique, an electro-polished specimen is illuminated by plane-polarized light, and the reflected light is passed through an analyzer with the polarization plane rotated by 90°. If the specimen surface is optically isotropic, the image will be dark. If, however, there are optically anisotropic (e.g., having a noncubic lattice) *crystallites* or surface films, they will be bright and colored because they rotate the polarization plane of the reflected light. Grains of the same phase can be discriminated because the previously mentioned rotation depends on their orientation.

pole   In crystallography, an intersection of the normal to any *lattice plane* of a crystal with a *reference sphere*; the crystal is assumed small and placed at the center of the sphere. The position of the pole, projected onto the *Wulff net*, characterizes the orientation of the lattice plane.

pole figure   *Stereographic projection* displaying the density of certain *poles* of all the *crystallites* on the *Wulff net*, definitively oriented with respect to the *polycrystalline* sample. If the density is inhomogeneous, then there is a *texture* in the sample. The indices of the poles used are commonly included in the pole figure name as, e.g., a {002} pole figure.

polychromatic [x-ray] radiation   See *white radiation*.

polycrystal   *Polycrystalline* body.

polycrystalline   Consisting of many *crystallites*.

polygonization   Formation and growth of *subgrains* usually observed at the last stage of *recovery*. Subgrains form from the dislocation cells (see *cell structure*) and their boundaries from *geometrically necessary dislocations* in the cell walls, whereas the other dislocations in the walls rearrange and annihilate. Subgrain growth proceeds in the same manner as *grain growth*. Commonly, polygonization results in the occurrence of *recrystallization nuclei*. However, sometimes, it can be the last stage of *substructure* alterations upon heating deformed materials, as, e.g., in the case of *continuous recrystallization*.

polygonized   Formed by *polygonization*.

polymorphic crystallization   Occurrence of a *crystalline phase* of the same *composition* as the parent *glassy phase*. See *devitrification*.

polymorphic modification   See *polymorphism*.

polymorphic transformation   *Spontaneous* transition of one *polymorphic modification* into another. It can evolve due to changes in temperature (as, e.g., *quartz* ↔ *tridymite* in *silica*, *austenite* ↔ *ferrite* in *steels*, or β-*phase* ↔ α-*phase* in Ti *alloys*) or pressure (e.g., *zinc blende structure* ↔ *rock salt structure* in CdTe). Polymorphic transformation is a *first-order transition*. *Martensitic transformation* can also be considered polymorphic, although in many cases, as, e.g., in steels or Ti alloys, martensite is a *metastable phase*. See *polymorphism*.

polymorphism   Existence of different *crystal structures* of the same *composition* (they are called polymorphic modifications) in certain temperature or pressure ranges in a solid substance. In *solid solutions*, polymorphism is connected with *allotropy* of its *components*, allotropy being a particular case of polymorphism. Compare with *polytypism*.

polytypism Existence of various *crystal structures* (known as polytypes) in an *intermediate phase*. Like *polymorphic modifications*, polytypes have different *crystal lattices*, but, in contrast to polymorphic forms, their lattices differ only in the stacking sequence of identical atomic layers. This is the reason why cubic and hexagonal polymorphic forms of ZnS, *sphalerite* and *wurzite*, can also be considered polytypes.

porosity. Volume fraction of pores. In *sintered* products, porosity is usually characterized by the actual density of a product divided by its theoretical density.

Portevin–Le Chatelier effect. See *dynamic strain aging*.

postdynamic recrystallization. Occurrence and growth of *strain*-free grains in the course of *annealing* after *hot deformation* is complete. In contrast to *metadynamic recrystallization*, postdynamic recrystallization starts in deformed materials in which *recrystallization nuclei* are lacking.

powder method. X-ray diffraction technique using *monochromatic radiation* wherein the sample is either a *fine-grained* non-*textured polycrystal* or a fine powder. Powder method is usually used for *x-ray structure analysis*.

powder pattern. X-ray diffraction pattern obtained by the *powder method*.

power-law creep. Stress dependence of the strain rate, $\dot{\varepsilon}$, at the steady-state dislocation creep in tension tests:

$$\dot{\varepsilon} \propto (D/kT)\sigma^n$$

where D is the *coefficient* of *self-diffusion* (at temperatures $T < 0.4T_m$ it is the *dislocation core* or *grain boundary self-diffusion* that controls the creep process, and at $T > 0.4T_m$ is the *bulk diffusion*), k is the *Boltzmann constant*, and σ is the *tensile stress*. In the case of *dislocation creep*, the exponent $n = 3$–5 at $\sigma/E < 10^{-4}$, where E is *Young's modulus*; however, in *dispersion-strengthened* materials, it is greater than 7. *Diffusional* and *Harper–Dorn creep* developing at low stresses ($\sigma/E < 5\,10^{-6}$) are described by the same law with the exponent $n = 1$.

precipitate. Particle of a new *phase* occurring upon *precipitation*. In the case of precipitation from a *solid solution*, the shape of precipitates is determined by *interfacial energy*, provided the *elastic strain energy* associated with the particles is not significant. As a result of *heterogeneous nucleation*, precipitates are usually distributed inhomogeneously over the *matrix phase*.

precipitated phase. Generally, a new phase occurring *upon precipitation*. In cases in which the parent phase is a *solid solution*, the precipitated phase is called the *second phase*. If the second phase is *stable*, the *matrix* phase should be a *saturated solid solution*. In the opposite case, i.e., at a *metastable* second phase, solid solution remains supersaturated, but to a smaller extent than before the precipitation (see Figure P.2).

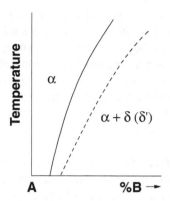

FIGURE P.2 Solvus lines for α solid solution. Solid line shows solvus at equilibrium with stable phase δ and dashed line shows solvus at equilibrium with metastable phase δ'.

precipitate reversion See *reversion*.

precipitation In a general sense, any *phase transition* evolving by the *nucleation* and growth of a new *phase*. In a restricted sense, it is a phase transition consisting in the occurrence of finely dispersed particles of a new phase from a *supersaturated solid solution*. In this case, the volume fraction of the new phase is relatively small.

precipitation-free zone (PFZ) Zone along *grain boundaries* where the amount of *precipitates* is reduced in comparison to the grain interior. It is the result of a decreased *diffusivity* of *substitutional solutes* in the vicinity of the boundaries. This, in turn, is a result of the depletion of the *quenched-in vacancies* disappearing at the grain boundaries (see *vacancy sink*).

precipitation strengthening/hardening Increase in the *flow stress* upon *aging treatment*, resulting from the hinderance of the *dislocation glide* motion by: *coherency strains* around the *precipitates* (see *coherency strain hardening*) and the necessity of cutting the *precipitates* (see *particle shearing*) or passing between them (see *Orowan mechanism*). Since the *dislocation density* in deformed precipitation-strengthened materials is much higher than in deformed single-*phase* materials, this can also lead to the following increment of the *tensile stress*, σ, at increased *tensile strains* ε:

$$\Delta\sigma \propto G \, (\varepsilon \, f/d)^{1/2}$$

where G is the *shear modulus*, and f and d are the volume fraction and the mean diameter of precipitates, respectively. Precipitation strengthening is also called age hardening.

precipitation treatment *Heat treatment* of a material subjected prior to *solution treatment*. Precipitation treatment is accompanied by a noticeable *precipitation strengthening* because of the occurrence of *GP zones* or *second-phase particles*.

preferred grain orientation See *crystallographic texture*.

preformed nucleus Nucleus that exists before the conditions necessary for its growth are attained, as, e.g., *recrystallization nucleus*.

preprecipitation Occurrence of *GP zones* preceding the new phase *precipitation*. See also *aging*.

primary α-phase [in Ti alloys] *Microconstituent* formed by *α-phase* grains present at the temperatures of the (α + β)-phase field. The same name is applied to the α-phase grains occurring at the boundaries of β-phase grains upon slow cooling from the temperature of the β-phase field. Primary α-phase grains should not be confused with *primary crystals* forming in the melt during *solidification*.

primary creep See *transient creep*.

primary crystals Single-*phase microconstituent* consisting of *grains* crystallized directly from the melt. Primary crystals can be *equiaxed* or *dendritic*, sometimes they have flat facets.

primary dislocation The same as *lattice dislocation* (in distinction to *grain boundary dislocation*, called *secondary dislocation*).

primary extinction X-ray *extinction* in nearly perfect crystals resulting from a partial canceling of the incident beam due to its interference with reflected beams inside a *crystal*.

primary recrystallization *Nucleation* and growth of new *strain*-free *grains* (see *recrystallization nucleus*) in *deformed matrix* upon *annealing plastically deformed* materials. The *driving force* for primary recrystallization (PR) is a decrease in the *elastic strain energy* proportional to the *dislocation density* or, in the case of well-developed *recovery*, a decrease in the energy associated with *subboundaries*. PR *kinetics* are usually described by the *Avrami equation*. PR is static, if it evolves upon annealing after the deformation completion; it is dynamic if it evolves simultaneously with *hot deformation*. PR can develop after the completion of hot deformation as, e.g., *metadynamic recrystallization* or *postdynamic recrystallization*. PR is also referred to as discontinuous recrystallization.

primary slip system First slip system to become active because the *critical resolved shear stress* on the system is attained earlier than on the others.

primary solid solution See *terminal solid solution*.

primary structure *Microstructure* formed upon *solidification*.

primitive lattice *Bravais lattice* wherein the *lattice points* are placed only at the vertices of its *unit cell*. It is also called simple lattice.

prismatic [dislocation] loop Closed loop of an *edge dislocation* that cannot *glide* over its plane because its *Burgers vector* is perpendicular to the plane. Prismatic loops can occur due to the collapse of the discs of *point defects* (see *Frank partial dislocation*) or from *dislocation dipoles* torn loose from *screw dislocations*.

prismatic slip *Slip* over a *prism plane* observed in *HCP alloys* with *c/a* < 1.633.

prism plane In *hexagonal lattices*, any plane perpendicular to the *basal plane*. There can be discerned first-order prism planes { $10\bar{1}0$ }, second-order ones { $11\bar{2}0$ }, etc.

proeutectoid Forming before an *eutectoid reaction*.

proeutectoid cementite *Microconstituent* formed by *cementite* precipitating from *austenite* in *hypereutectoid steels* upon their cooling below A_{cm}, i.e., due to a decrease of the carbon solubility in austenite according to the ES line in Fe–Fe$_3$C diagram. At a slow cooling, proeutectoid cementite precipitates at the austenite *grain boundaries* and forms a *carbide network*. At an increased cooling rate, it precipitates inside the austenite *grains* and forms *Widmannstätten structure*. Proeutectoid cementite is also called secondary cementite.

proeutectoid ferrite *Microconstituent* formed by *ferrite* precipitating from *austenite* in *hypoeutectoid steels* upon their cooling below A_3. Upon slow cooling, proeutectoid ferrite occurs at the *austenite grain boundaries*, and its grains are *equiaxed* or form a network, depending on their volume fraction. At an increased cooling rate, especially when austenite is *coarse-grained*, crystallites of proeutectoid ferrite have a lath- or plate-like shape typical of the *Widmannstätten structure* (see *Widmannstätten ferrite*).

pseudoplasticity In *alloys* with a slowly growing *martensite*, an occurrence of a noticeable *plastic deformation* (up to *tensile strains* of ~5%) upon loading an object in the range between the M_d and M_f temperatures, i.e., in the course of *martensitic transformation*. This strain is caused by the growth of martensite crystallites whose orientation is selected by the applied force in such a way that they produce a *shear* best compatible with the force. See also *one-way shape memory effect*.

pyramidal plane In *hexagonal lattices*, any plane inclined to a *basal plane* at an angle <90°, e.g., a first-order pyramidal plane { 10$\bar{1}$1 } or a second-order one { 11$\bar{2}$2 }.

pyramidal slip *Slip* over a *pyramidal plane* observed in some *HCP alloys*.

Q

quantitative metallography Complex of methods for determining certain quantitative *microstructural* parameters, e.g., the *mean size* and the shape of *grains, grain size distribution*, volume fraction of *disperse particles* and other *microconstituents*, interparticle spacing, etc. Since the *metallographic samples* are opaque, all these data relate to two-dimensional sections of a three-dimensional *structure*. Thus, they are to be converted into three-dimensional parameters using image analysis. Conversion methods are given in quantitative metallography that is also referred to as stereology.

quartz Low-temperature *polymorphic modification* of *silica* known as "high," or β-, quartz. It has a *hexagonal crystal structure* and, upon cooling at an increased rate, transforms at 573°C into a *metastable* "low," or α-, quartz with a *trigonal* structure, the transformation being *displacive* and accompanied by significant volume changes.

quasi-crystal *Solid phase* with *atomic structure* characterized by, e.g., a 20-fold rotational *symmetry* incompatible with the periodic structure of *crystal lattices* (see *n-fold axis*). This kind of macroscopic symmetry is typical of polyhedra that cannot fill the space as perfectly as *unit cells* can.

quasi-isotropic See *isotropic*.

quench aging Evolution of fine *precipitates* from a *supersaturated solid solution*. This term is used for low-carbon *steels*; in other *alloys*, the same phenomenon is referred to as *aging* or *precipitation*.

quench hardening See *precipitation hardening*.

quench-in vacancies Excess vacancies whose *concentration* at a given temperature exceeds the equilibrium concentration, due to a rapid cooling from high temperatures. These vacancies tend to migrate to *vacancy sinks*, which increases the *diffusion* rate, accelerates the *nucleation* and growth of the *GP zones* and *precipitates*, and results in the occurrence of *segregations* and *precipitation-free zones* at *grain boundaries*.

quenching *Heat treatment* aimed at receiving *martensite*. It comprises heating to a definite temperature, immediately followed by cooling at a *critical rate* down to temperatures below M_f. *Hypoeutectoid* and *hypereutectoid* steels are heated, respectively, to temperatures of the γ-field and (γ + Fe$_3$C)-phase field in Fe–Fe$_3$C diagram. After quenching, the *microstructure* of the steels consists of martensite or martensite and cementite, respectively. Some low-carbon steels are quenched from temperatures of the $(\alpha + \gamma)$-field to produce a *dual-phase microstructure* consisting of *ferrite* and martensite. Titanium *alloys* are quenched primarily from temperatures of the $(\alpha + \beta)$-phase field in the corresponding *phase diagram*. If quenching is interrupted between M_s and M_f temperatures, a *retained austenite* (in steels) or *retained β-phase* (in Ti alloys) is present along with martensite. Quenching is also called hardening treatment.

R

radial distribution function (RDF) Averaged atomic density vs. the distance from an atom chosen as the origin. RDF is used for describing the *atomic structure* of *glassy phases*.

radiation damage See *irradiation damage*.

random grain boundary See *general grain boundary*.

random solution See *disordered solid solution*.

R-center *Color center* occurring due to the trapping of two electrons by an agglomerate of two vacancies in the cation *sublattice*.

reciprocal lattice Geometric construction used for resolving various diffraction problems (see *Ewald sphere*). It is obtained as follows. Take vectors starting from one origin and directed perpendicular to $\{hkl\}$ planes of the *crystal lattice* considered. The vector lengths are $1/d_{hkl}$, where d_{hkl} is the distance between $\{hkl\}$ planes. The sites at the ends of the vectors form a reciprocal lattice.

reconstructive transformation The name of *diffusional transformation* used in ceramic science.

recovery *Substructure* changes evolving upon heating a *plastically deformed* body and not connected with the migration of *high-angle boundaries*. Recovery (R) comprises several stages, including the disappearance of deformation-induced *point defects* and *annihilation* and rearrangement of *dislocations* (by both *glide* and *climb*), as well as *subgrain* formation and growth (i.e., *polygonization*). Usually, R precedes *primary recrystallization* and can develop in the deformed matrix simultaneously with the formation of *recrystallization nuclei*. However, R can also be a sole process, decreasing the concentration of *crystal defects*, as in the case of *continuous recrystallization*. R is called static if it evolves upon annealing after the deformation completion, or dynamic if it evolves simultaneously with the plastic deformation (usually in the course of *hot deformation*). Restoration of physical or mechanical properties accompanying *annealing treatment* of plastically deformed materials is also termed recovery, as, e.g., the recovery of electrical conductivity.

recrystallization (ReX) Usually, *primary recrystallization*. The same term is used to describe the occurrence and growth of *crystalline phases* from *amorphous* phases upon heating, i.e., *crystallization* or, in the case of *glasses*, *devitrification*.

recrystallization annealing *Heat treatment* of *plastically deformed* bodies, with the primary aim being to delete completely the *work hardening* by heating above the *recrystallization temperature*. It results in *primary recrystallization* and *grain growth*, usually *normal*, but sometimes *abnormal*. Recrystallization annealing is accompanied by formation of an *annealing texture*. The *mean grain size* after recrystallization annealing depends upon both the material purity and the initial *microstructure* and on the annealing conditions, as well as on the type and rate of deformation, and is often depicted in a *recrystallization diagram*.

recrystallization diagram Dependence of the *mean grain size* after *recrystallization annealing* versus *strain* in preceding *plastic deformation*. The temperature and duration of the annealing are the same for all the strains. Sometimes, a three-dimensional diagram is used in which the annealing temperature is shown along the third axis (in this case, the annealing duration only is constant). Since *recrystallization temperature* is different for different strains, whereas the annealing temperature is the same, the extent of the *grain growth* evolution upon annealing at identical temperatures after different strains is different.

recrystallization *in situ* See *continuous recrystallization*.

recrystallization nucleus In *plastically deformed* materials, an area characterized by low *dislocation density*, as, for example, in a *strain*-free material, and separated from its surroundings by a *high-angle boundary* of *high mobility*. Recrystallization nuclei (RN) occur upon *recrystallization annealing*, although after severe *cold deformation*, *preformed* recrystallization nuclei are also present. The RN formation proceeds mainly by *subgrain* growth (analogous to *grain growth*), in the course of which some subgrains acquire an increased *disorientation* with respect to their neighbors. Because of this, recrystallization nuclei occur in the matrix areas characterized by increased orientation gradients, e.g., near the boundaries of deformed grains, in *deformation bands* and *shear bands*, at coarse hard inclusions (see *particle-stimulated nucleation*), etc., but not in the *matrix bands*. An increase of subgrain disorientation is accompanied by an increase of the mobility of the subgrain boundary, and a significant mobility rise corresponds to the transformation of the *subboundary* into a high-angle boundary. As follows from thermodynamics, a subgrain with a high-angle boundary can survive if it has an increased size in comparison to the adjacent subgrains. It is also possible that large subgrains emerge by *subgrain coalescence*. Since subgrain growth is *diffusion-controlled*, *thermal activation* is necessary for recrystallization nuclei formation. This is the reason why *primary recrystallization* starts at an increased temperature (see *recrystallization temperature*).

recrystallization temperature Minimum temperature of an *annealing* treatment resulting in the onset of *primary recrystallization*. Recrystallization temperature (RT) decreases with the deformation degree and remains approximately constant at high *strains*. At high strains, RT is ~0.4 T_m in materials of commercial purity, but can be much lower (~0.2 T_m and lower) in high-purity materials. Furthermore, RT depends on many parameters affecting the *substructure* of *plastically deformed* materials and the occurrence of *recrystallization nuclei*, such as deformation rate, initial grain *size* and *texture*, presence of *precipitated particles*, etc. Many *solutes* increase RT, especially at low *concentrations*. The effect of *precipitates* on RT depends on the mean interparticle spacing, Λ: uniformly distributed particles with a small Λ increase RT by retarding the dislocation rearrangements as well as the formation and migration of *subboundaries*, whereas coarse particles with an increased Λ reduce RT (see *particle-stimulated nucleation*). In addition, RT can be greatly affected by the preceding *recovery* and depends on both the heating rate and the annealing duration.

recrystallization texture *Crystallographic texture* forming in the course of *primary recrystallization*; it is commonly observed in articles with a noticeable *deformation texture*. Recrystallization texture develops via growth competition of *recrystallization nuclei* whose orientations are usually not random.

recrystallized [Structure or grain] formed as a result of *primary recrystallization*, either *static* or *dynamic*.

reference sphere See *stereographic projection*.

reflection high-energy electron diffraction (RHEED) Technique for studying *crystal lattice* of single-crystalline samples using a glancing primary beam of high-energy electrons. Because of the higher energy of primary electrons in comparison to *LEED*, RHEED enables the discovery of in-plane *lattice constants* in layers as thin as ~5 *atomic radii* by means of *electron diffraction patterns*.

reflection sphere See *Ewald sphere*.

relaxation modulus See *anelasticity*.

relaxation time Period of time over which the initial magnitude of a certain quantity changes by e times ($e \cong 2.718$), provided the other operating factors remain unchanged. For instance, in the case of static loading, *anelasticity* is revealed in an *elastic strain*, increasing with time, whereas both the *stress* and temperature are constant.

replica Imprint from the surface of a polished and slightly etched sample. Replicas are obtained by the deposition of an amorphous material onto the sample surface, followed by separation of the deposited layer. The layer copies the topography of the sample surface and thus has a varying thickness. There can be direct (i.e., one-stage) and indirect (two-stage) replicas, the latter being the imprints from direct replicas. There are also *extraction replicas*.

residual austenite See *retained austenite*.

residual electrical resistance Magnitude of electrical resistance of a given specimen at 4 K, R_{4K}. Since R_{4K} of nondeformed metals depends mainly on the *solute* content (see *Matthiessen's rule*), the ratio R_{300K}/R_{4K} (where R_{300K}

is the electrical resistance of the same specimen at the room temperature) is used as a measure of the solute *concentration*: the higher the concentration, the lower the ratio.

residual stresses Balanced *stresses* remaining in a body after the removal of external actions. Residual stresses induce *elastic deformation*, which can cause *plastic deformation* (see *stress relaxation*) and affect *vacancy* flux (see *diffusional plasticity*), dislocation *glide* motion (see, e.g., *strain hardening*, *Orowan mechanism*, and *coherency strain hardening*), and evolution of *phase transitions* (see, e.g., *stress-assisted* and *strain-induced martensite*, as well as *transformation induced plasticity*). In some cases, residual stresses result in the occurrence of cracks. Residual stresses are also known as internal stresses. See *thermal stress* and *macroscopic* and *microscopic stresses*.

resolution limit Minimum distance between two points or parallel lines in an object that can be seen separately. Resolution limit, d_{res}, of a lens depends upon its *numerical aperture*, A_N, and the light wavelength, λ:

$$d_{res} \cong 0.5\lambda/A_N$$

It characterizes quantitatively the *resolving power* of the lens.

resolved shear stress Component of shear stress acting on a chosen *slip plane* along a certain coplanar *slip direction*.

resolving power Quantity reciprocal to the *resolution limit*.

retained austenite *Austenite* remaining untransformed upon *quenching* below M_s temperature, but above M_f. If the carbon *diffusivity* is high enough, *carbides* can precipitate from retained austenite, and it transforms into *secondary martensite*. Retained austenite is also called residual austenite.

retained β-phase In Ti alloys, β-*phase* remaining untransformed upon *quenching* below M_s temperature. *Aging treatment* after quenching results in *decomposition* of retained β-phase (see *aging in Ti alloys*).

retrograde solidus *Solidus* configuration corresponding upon temperature lowering first to an increase of the *concentration* of a *component* dissolved in a *solid phase*, and then to its decrease. Commonly, a decrease in temperature results in an increasing solubility in the corresponding solid phase.

reversibility In materials science, the sequence of *phase transitions* observed upon reheating that is identical to that evolving upon cooling, but in the reverse order. Reversibility is always observed when *equilibrium phases* are involved and the reheating temperature exceeds the equilibrium temperature of the high-temperature phase, as, e.g., during *eutectoid reaction*. If, however, at least one of the phases is *metastable*, reversibility is not observed, as, e.g., on *tempering* of *steel martensite*. At the same time, reversibility, not only of the *phase composition*, but also of the *microstructure* can be observed in the course of the backward *martensitic transformation* in some nonferrous *alloys* heated above the A_f temperature (see *shape memory effect*). In irreversible Fe–Ni alloys, a significant *thermal hysteresis* of the martensitic $\gamma \leftrightarrow \alpha$ transition is observed, i.e., M_s is significantly lower than A_s. At the same temperature between A_s and M_s, such an alloy consists of γ-phase upon cooling and of α-phase upon heating.

reversible temper brittleness Embrittlement in *low alloy steels* observed upon slow cooling after *tempering treatment* at 450–550°C. It can be removed by reheating to 600–650°C and a subsequent rapid cooling. This kind of brittleness is induced by *equilibrium segregation* of definite impurities to the *grain boundaries*. The reheating results in impurity desegregation, and the rapid cooling prevents the formation of new segregations.

reversion In the course of *aging treatment*, dissolution of *metastable phase* preceding the occurrence of a new phase, more stable than the first one (see *rule of stages*). It is also referred to as precipitate reversion.

rhombohedral system See *trigonal system.*

rocking curve *Diffraction* curve obtained as follows. The primary x-ray beam and the counter are fixed at a *diffraction angle* 2θ, corresponding to a certain diffracting plane, whereas the sample mounted at the *diffractometer* is rotated about the diffractometer axis in an angular range $\theta \pm \Delta\theta$. The width of the rocking curve can be used for determining either the *scatter* of an *axial texture* (in thin films) or the quality of semiconductor *single-crystals.*

rock salt [structure] type *Crystal structure* typical of many *ionic crystals*; it is identical to the *NaCl structure.*

Rodrigues vector Description of grain *disorientation* in a *polycrystal* by the vector:

$$\mathbf{R} = \mathbf{n} \tan \Theta/2$$

where Θ is the *disorientation angle*, and $\mathbf{n}$ is the vector of the corresponding disorientation axis. A disorientation function can be described by the Rodrigues-vector distribution in what is known as Rodrigues space, whose coordinate system is the same as that of the *crystal lattice* concerned.

roller quenching See *melt spinning.*

R-orientation One of the main *components*, $\{124\}\langle211\rangle$, of *annealing texture* in cold-rolled *FCC* metals.

rotating crystal method Technique of the *x-ray structure analysis* wherein a needle-like *single crystal*, rotated around its long axis, is irradiated by a *polychromatic* x-ray beam directed perpendicular to the axis. A diffraction pattern is registered on a cylindrical film (or screen), the cylinder axis being coincident with the crystal's long axis.

rule of stages According to this rule, *precipitates* of a *metastable phase* or *GP zones* dissolve in a *supersaturated solid solution* upon heating, and a new, stable or more stable phase forms (the phenomenon is known as reversion). This is connected with different *solubility limits* for different precipitating phases and GP zones, the metastable precipitates having a greater *solubility* than the more-stable ones (see Figure R.1).

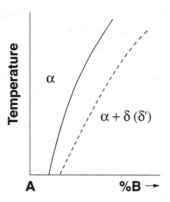

FIGURE R.1 Solubility limits for stable phase δ (solid line) and metastable phase δ′ (dashed line). For further details, see *precipitated phase*, Figure P.2.

Because of this, heating above the *solvus* of any metastable phase is accompanied by its dissolution, which makes possible precipitation of a more stable phase.

r-value Ratio of the *true strains* in the longitudinal and transverse directions in a flat *tension* sample, the longitudinal direction being parallel to *RD* of a sheet from which the sample was cut out. The strains are determined at a definite (usually 15–20%) homogeneous *nominal strain*. The r-value is also referred to as strain ratio.

r̄-value Average of *r-values* found on *tension* specimens cut out of a sheet at different angles to *RD*:

$$\bar{r} = (r_0 + r_{90} + 2r_{45})/4$$

(subscripts designate the angles between the specimen axis and RD). The r̄-value characterizes *normal anisotropy*; it is also referred to as Lankford coefficient.

S

σ-plot Dependence of the *interfacial energy*, σ, on the *interface* orientation in three-dimensional polar coordinates. Two-dimensional sections of σ-plot are usually considered. Cusps in σ-plot correspond to the *coherent* or *partially coherent interface.*

60°-dislocation *Mixed perfect dislocation* whose *Burgers vector* lies at 60° to the dislocation line. It is usually observed, along with *screw* dislocations, in *diamond* and *zinc blende structures.*

Sachs factor Quantity averaging the influence of various grain orientations on the *resolved shear stress*, τ_r, in a *polycrystal*:

$$\sigma = M\tau_r$$

where M is the Sachs factor and σ is the *flow stress*. The averaging is undertaken under the supposition that each *grain* deforms separately. Reciprocal Sachs factor can be used for polycrystals, instead of *Schmid factor*, whose magnitude is defined for a single grain only. Reciprocal Sachs factor is 0.446 in a nontextured polycrystal with *FCC structure*.

sample thickness effect *Grain growth* inhibition observed in thin sheets and films: the mean grain diameter in these objects does not exceed $\sim 2\delta$, where δ is the sheet (film) thickness. It is assumed that the effect is a result of the *groove drag.*

saturated solid solution Solid solution whose *composition* corresponds to *solubility limit* at the temperature concerned.

S/Bs-orientation One of the main *texture components*, $\{168\}\langle 211 \rangle$, observed in cold-rolled *FCC* metallic materials of low *stacking-fault energy.*

scanning Auger-electron microscope (SAM) *SEM* combined with *AES*; it is used for simultaneous analysis of the surface layer chemistry and *microstructure.*

scanning electron microscope (SEM) Device for studying the surface topography, *microstructure*, and chemistry of metallic and nonmetallic specimens at *magnifications* from 50 up to $\sim 100{,}000\times$, with a *resolution limit* <10 nm (down to ~ 1 nm) and a *depth of focus* up to several μm (at magnifications $\sim 10{,}000\times$). In SEM, a specimen is irradiated by an electron beam and data on the specimen are delivered by *secondary electrons* coming from the surface layer of thickness ~5 nm and by *backscattered electrons* emitted from the volume of linear size ~0.5 µm. Due to its high depth of focus, SEM is frequently used for studying fracture surfaces. High *resolving power* makes SEM quite useful in *metallographic examinations*. Sensibility of backscattered electrons to the atomic number is used for the detection of *phases* of different chemistry (see *EPMA*). *Electron channeling* in SEM makes it possible to find the orientation of *single crystals* by *ECP* or of *grains* by *SACP* (see also *orientation imaging microscopy*).

scanning transmission electron microscope (STEM) *TEM* wherein a fine *electron probe* is scanned over the specimen surface, and the chemistry of the specimen is studied using transmitted electrons (*EELS*) or emitted x-rays (*EDS*). STEM gives images of the specimen's *substructure* with a lateral *resolution* better than 1 nm.

scanning tunneling microscope (STM) Device for studying the surface topography of solid electronic conductors with a lateral *resolution* better than the *atomic size*. In STM, a sharp microscope tip is scanned over the specimen surface without touching it, and at the same time, the tunneling current between the tip and the surface atoms, proportional to the distance between them, is recorded. The results obtained are transformed into the images displaying the *atomic structure* of a clean surface or the *adatom* arrangement.

Schmid factor Quantity

$$m_s = \tau_r/\sigma$$

determining the *resolved shear stress*, τ_r, in different *slip systems* of a *grain* (*single crystal*) under the influence of *tensile stress* σ:

$$m_s = \cos \phi \cos \psi$$

Here, ϕ is the angle between the force and the normal to the *slip plane*, and ψ is the angle between the force and the *slip direction*. The maximum $m_s = 0.5$ is reached on the slip systems with $\phi = \psi = 45°$. See also *Sachs factor* and *Taylor factor*.

Schmid's law Empirical law, according to which, the *slip* in *single crystals* starts at the same *resolved shear stress* on any *slip system* if the stress magnitude exceeds a critical value for a particular material (see *Peierls stress*).

Schottky pair Complex of two *point defects* consisting of cation and anion *vacancies* and observed in *ionic crystals*. See also *structural disorder*.

screw dislocation Dislocation whose *Burgers vector* is parallel to its line. Screw dislocation converts the family of parallel lattice planes, normal to the dislocation line, into a continuous screw surface, the dislocation being its axis. Since screw surfaces can be right-hand and left-hand, there can be the right-handed and left-handed screw dislocations. It should be noted that the latter depends on the choice of the Burgers vector direction and thus is a conventional characteristic. See also *dislocation sense*.

S crystal See *smectic crystal*.

secondary cementite See *proeutectoid cementite*.

secondary creep See *steady-state creep*.

secondary crystals *Microconstituent* occurring in the solid state due to the evolution of a new *phase* caused by a decrease of the *solubility limit* upon the temperature decreasing. The term is used for to describe *microstructure* where there are *primary crystals* of the same phase. See also *second phase*.

secondary dislocation See *grain-boundary dislocation*.

secondary electron Electron emitted under the influence of the primary x-ray (or electron) beam resulting from *inelastic scattering* of the primary radiation.

secondary extinction Decrease in the intensity of a reflected x-ray beam due to the shielding of the incident beam by *subgrains* in the sample surface layer that reflect primary radiation. Secondary extinction can be observed not only in nearly perfect *crystals* (i.e., together with *primary extinction*), but also in *polycrystalline* and *single-crystalline* samples with clearly developed *substructure*.

secondary hardening *Hardness* increase in some *alloy steels* caused by *precipitation* of *special carbide*s upon *tempering treatment*.

secondary ion mass spectroscopy (SIMS) Technique for elemental chemical analysis wherein the primary ion beam (ion energy of 1–10 keV) produces secondary ions that are analyzed by mass spectroscopy. SIMS can supply data from quite a small area through focusing the primary ion beam to a diameter smaller than 50 nm. Since the primary beam *sputters* the surface layer of the specimen, SIMS can be used for a layer-by-layer analysis with a *resolution* from 1 to 5 *atomic sizes*.

secondary precipitate See *second phase*.

secondary recrystallization See *abnormal grain growth*.

secondary slip system Any slip system except the *primary* one.

secondary structure *Microstructure* forming after the *solidification* is complete and resulting from *phase transitions* in the solid state.

second-order transition *Phase transition* accompanied by continuous changes in *free energy* at the transition temperature. At these transitions, the new phase cannot coexist with the parent phase and the latter cannot be *supercooled* or *superheated*. See, e.g., *magnetic transformation*. Compare with *first-order transition*.

second-order twin Twin inside a *first-order twin*.

second phase *Phase constituent* evolving by *precipitation* from a *solid solution*, its volume fraction usually being not greater than ~10%. The second phase *crystallites* form a *microconstituent* sometimes referred to as *secondary crystals*.

seed crystal Small *single crystal* used as *preformed nucleus* for growing a larger single crystal from the *melt*.

segregation Increased *concentration* of *solutes*, both *alloying elements* and *impurities*, in certain areas of an article or ingot (see *macrosegregation*), inside *grains* (see *coring* and *cellular microsegregation*), or at certain *crystal defects* (see *grain-boundary segregation, Cottrell atmosphere*, etc.) in comparison to their average concentration. See also *equilibrium segregation* and *nonequilibrium segregation*.

selected area channeling pattern (SACP) See *electron channeling*.

selected area diffraction (SAD/ESAD) In *TEM, electron diffraction* technique for *crystal lattice* analysis of small *grains* of size ~1 μm at the electron energy 100 keV and ~10 nm at 1 MeV. SAD patterns are also used for determining the *subgrain* (grain) orientation, as well as the *orientation relationship* between grains of different *phases*.

self-diffusion *Diffusion* of *host* atoms (see *vacancy mechanism*).

self-interstitial *Host atom* occupying an *interstitial void*. In contrast to *vacancies*, the equilibrium *concentration* of self-interstitials is negligible because the *free energy* of their formation is quite high in comparison to vacancies. They occur primarily as a result of the specimen irradiation with high-energy particles and can exist as individual defects. Self-interstitial is sometimes called interstitialcy.

self-similar Retaining its shape with time, provided the scale transformation is taken into account. For instance, the *grain size distribution* in the course of *normal grain growth* is self-similar, which can be proved using a reduced grain size, D_i/D_M, instead of D_i (D_i is the grain size corresponding to the *i*-th size interval, and D_M is the *most probable size*).

semi-coherent interface See *partially coherent interface*.

sessile dislocation Partial dislocation that cannot *glide* because its *Burgers vector* has a component normal to its *slip plane*.

shadowing Deposition of metal vapor on a *replica* wherein the vapor flow is directed at an acute angle to the replica surface, with the aim to enhance the *absorption contrast*.

shallow impurity *Solute* or *impurity* atom whose energy level lies inside the *band gap* and is characterized by the activation energy comparable with kT, where k is the *Boltzmann constant,* and T is the absolute temperature. *Acceptors* and *donors* are typical shallow impurities. Some other *lattice defects*, e.g., *vacancies* in elemental semiconductors and *ionic crystals*, can also act as shallow impurities.

shape memory effect Reversibility of the specimen shape following temperature alterations. Such a behavior is observed in *alloys* undergoing *martensitic transformation* with a slow growth of martensite crystallites. Cooling of these alloys through the M_s–M_f temperature range produces martensite, and reheating through the A_s–A_f range restores the parent phase. Reversible shape changes take place in the case of martensite orientation according to only one of many possible variants of its *orientation relationship* with the parent phase. The variant selection is governed by *residual stresses* imposed by preliminary "training." The described shape memory effect is also referred to as the two-way shape memory effect. See also *reversibility, pseudoplasticity,* and *one-way shape memory effect.*

sharp yield point Appearance of both the *upper* and *lower yield stresses* on *stress–deformation diagram*. The stress drop, from the upper to the lower yield stress, can be explained by the occurrence of fresh *dislocations* able to *glide*. They are either new dislocations produced by *dislocations sources* or dislocations released from the *Cottrell atmospheres*. A release of the *elastic strain energy*, stored in the specimen before attaining the upper yield stress, leads to the dislocation glide motion and produces a *yield-point elongation*. Sharp yield point is usually observed in low-carbon *steels* with *interstitially* dissolved carbon or nitrogen, and is accompanied by the formation and evolution of *Lüders bands*. It can be avoided by alloying with elements that reduce the *concentration* of dissolved interstitial impurities. Sharp yield point can be temporarily suppressed by a small *plastic deformation* whose magnitude exceeds slightly the yield-point elongation (see *temper rolling*).

shear In material science, a loading scheme wherein two pairs of equal and opposite *shear stresses*, τ, act as shown in Figure S.1. They deform a rectangular parallelepiped in such a way that it becomes oblique. *Shear-type transformations* in the solid state are not connected with the action of shear stresses, although the latter can affect the transformations (see, e.g., *strain-induced martensite* and *pseudoplasticity*).

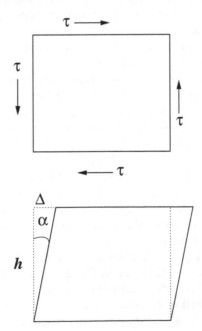

FIGURE S.1 Shear and shear strain. Above: before shear. Shear stresses τ are applied not only to horizontal faces of the parallelogram, but also to its side faces; the latter is necessary to prevent rotation of the parallelogram. Below: after shear. See text.

shear band Narrow band (of ~1 μm thickness) running across many *grains* in *plastically deformed* metal. Shear bands are observed in heavily *cold-rolled alloys* at *true strains* higher than 1. The cell size in shear bands is smaller and the cell walls are thinner than in the surrounding *matrix* (see *cell structure*). The *lattice* orientation within shear bands differs strongly from that of the matrix.

shear modulus *Elastic modulus* at *shear* deformation. In *polycrystals* without *texture*, it is *isotropic*. In the polycrystals, shear modulus *(G)* and *Young's modulus (E)* are connected by the relation:

$$G = E/2(1 + \nu)$$

where ν is the *Poisson's ratio*. Shear modulus is also known as modulus of rigidity. See *Hooke's law*.

shear strain *Strain* produced by *shear stresses* (see Figure S.1); it is determined as the ratio of the displacement, Δ, of the upper face of an initially rectangular parallelepiped to the distance, h, between the upper face and the other parallel face:

$$\gamma = \Delta/h = \tan \alpha$$

Since Δ is usually much smaller than h, it is assumed that $\gamma \cong \alpha$.

shear stress Stress induced by a force component acting parallel to a certain plane in a sample.

shear[-type] transformation *Phase transition* characterized by a *glissile* motion of the *interface* between the parent phase and a new one, the motion can be described as a *shear* transforming the *lattice* of the parent phase into the lattice of the new phase. In the course of the transformation, the atomic displacements are coupled and their magnitude is smaller than the *interatomic spacing*. Because of this, each atom in the new phase retains the same nearest neighbors as in the parent phase. New phase *nuclei* appear without an *incubation period,* and the volume fraction of the transformation products changes very little with time, if at all (see *athermal transformation*). Since shear transformation is always *nondiffusional*, the *composition* of the new phase is the same as that of the parent phase. Shear-type transformation is also called displacive transformation.

sheet texture *Preferred orientation* wherein a *lattice direction* [*uvw*] in the majority of *grains* is oriented along *RD*, and a certain plane of *zone* [*uvw*] is oriented parallel to the sheet (strip) surface.

Shockley partial dislocation *Glissile partial dislocation* in *FCC lattice* formed via *splitting* of a *perfect dislocation*. There are always two Shockley partials and a narrow ribbon of *stacking fault* between them. Shockley dislocation can *glide* over a single plane only. *Cross-slip* of Shockley dislocations is hindered because beforehand they should be transformed into a perfect dislocation, which, in turn, requires an additional energy.

Nevertheless, in materials with low *stacking-fault energy*, cross-slip is observed under the influence of *thermal activation*.

short-circuit diffusion path Diffusion path coinciding with *linear* (*dislocations* and *triple joints*) or *planar* (*grain* and *phase boundaries*) *crystal defects*. The activation energy for diffusion over these defects is much lower than for the *bulk diffusion*. The permeability of these defects for mass transport is determined by the extent of the *long-range order* distortions in their *atomic structure*. For instance, the *activation energy* for self-diffusion over a *coherent twin boundary* is close to that for *bulk self-diffusion*, whereas the activation energy for self-diffusion over an *incoherent* twin boundary or *general grain boundary* is ~2 times lower than for bulk self-diffusion. The *diffusvity* over short-circuit paths also depends on *solute segregation*.

short-range Relating to distances compared with an *interatomic spacing* in the solid state.

short-range order Characteristic of an *atomic structure* restricted to the nearest neighbors only; it is usually described by the number and type of neighbors. Short-range order is observed in both *crystalline* and *amorphous solids* (compare with *long-range order*). This is not to be confused with *short-range ordering* in *solid solutions*.

short-range ordering *Spontaneous* changes in the local arrangement of *solute* atoms in a *crystal lattice,* revealing itself in a tendency of the solute atoms to be surrounded by *host* atoms and vice versa. Short-range ordering can be observed above the temperature of the *order–disorder transformation* (see *short-range order parameter*).

short-range order parameter Quantity characterizing the arrangement of *solute* atoms over the sites of the first *coordination sphere* of a *host* atom in *solid solution*:

$$\sigma = (q - q_r)/(q_0 - q_r)$$

Here, q is the fraction of the solute-solvent atom pairs, q_r is the fraction of the pairs in a *random solid solution*, and q_0 is the fraction of the pairs in a perfectly *ordered solid solution* (see *order–disorder transformation*). It is seen that $\sigma \neq 0$ at $q_0 = 0$.

shrinkage Changes in shape and a decrease in the linear dimensions of a solid body. In *crystalline* materials, shrinkage occurs as a result of a reduction of the *specific volume* in the course of a *phase transformation* or due to a decrease of *porosity* during *sintering*.

sialon Oxynitride in the *system* Si–Al–O–N. For instance, α' and β' sialons are *solid solutions* of Al_2O_3 in α-Si_3N_4 and β-Si_3N_4, respectively, doped with Y_2O_3, MgO, BeO, etc, whereas O$'$ sialon is a solid solution of Al_2O_3 in the *intermediate phase* Si_2N_2O. In the sialon *lattice*, the sites of Si^{4+} ions are partially occupied by Al^{3+} ions and the sites of N^{3-} ions by O^{2-} ions. The amount of *dopants* in the sialons can be found with the aid of *equivalence diagram*.

silica Silicon dioxide SiO_2. Interatomic *bond* in silica is partially *covalent* and partially *ionic* (see *electronegativity*). It has three *polymorphic modifications*: *cristobalite*, *tridymite*, and *quartz*, with the transformation temperatures 1470 (cristobalite $\leftrightarrow$ tridymite) and 867°C (tridymite $\leftrightarrow$ quartz). In all of the modifications, Si atoms are arranged at the centers of tetrahedra formed by O atoms.

simple lattice See *primitive lattice*.

single crystal Body consisting of one crystal only. There are no *grain boundaries* in single crystals, although *subboundaries* and sometimes *twin boundaries* can be found.

single-domain particle Magnetic particle whose minimum linear size is smaller than the *domain wall* thickness; because of this, it consists of one *magnetic domain*. If several domains were present in such a particle, the particle's *free energy* would be increased. In the particle, the energy of the magnetic poles at its surface is the lowest in the case of the largest pole spacing. Thus, in a single-domain particle of an elongated shape, the orientation of its magnetization vector is determined not only by its *magnetic crystalline anisotropy*, but also by its shape anisotropy. If elongated single-domain particles are oriented predominately along the same direction in a body, the latter possesses a *magnetic texture* and excellent hard-magnetic properties.

single slip *Dislocation glide* motion over a single *slip system* characterized by the maximum *Schmid factor*.

sintering Procedure for manufacturing dense articles from porous particulate compacts (*porosity* in *green* compacts usually is between 25 and 50 vol%) resulting from *spontaneous* bonding of adjacent particles. The main *driving force* for sintering is a decrease of an excess *free energy* associated with the *phase boundaries*. Sintering is fulfilled by *firing* the compacts at high temperatures (up to ~0.9 T_m), and is always accompanied by their *shrinkage* and densification (i.e., a decrease in porosity). Shrinkage evolves primarily through *coalescence* of neighboring particles under the influence of the *capillary force* in the neck between the particles. The pore healing also contributes to shrinkage. Densification during sintering is accomplished by both the surface diffusion and the *grain-boundary diffusion*. It is essential for densification that the pores remain at the *grain boundaries*, because the pores inside the grains can be eliminated by slow *bulk diffusion* only, whereas the grain-boundary pores "dissolve," via the splitting out of *vacancies* and their motion to *sinks*, by much more rapid grain-boundary diffusion. Thus, the theoretical density can be achieved in cases in which the *abnormal grain growth* is suppressed and the rate of *normal grain growth* is low (for details of *microstructure* evolution in the course of sintering, see *solid-state sintering*). Sintering can be accelerated in the presence of a liquid phase (see *liquid-phase sintering*) or by pressure application during firing (see *hot pressing*).

size distribution Histogram displaying the frequency of *grains* (or particles) of different sizes. The shape of grain size distribution after *normal grain* *growth* is usually approximated by a curve close to lognormal, with the ratio of the maximum grain size, D_{max}, to the *most probable* one, D_M, close to ~3. However, in some cases, e.g., in the course of *abnormal grain growth*, grain size distribution has a significantly greater D_{max}/D_M ratio and can be described by a *bimodal* curve if the grain size distribution is determined by methods not taking into account grain volume fractions.

slip In materials science, a *mechanism* of *plastic deformation* wherein a part of a *crystallite* many *interatomic spacings* thick is translated on a plane (known as *slip plane*) relative to the remainder of the crystallite, i.e., all the *lattice points* in the former are moved over equal distances (see Figure S.2). Slip is produced by the *dislocation glide*, which explains why the terms "glide" and "slip" are frequently used as synonyms. Slip is accompanied by the lattice rotation in relation to the sample axes, which leads to the evolution of *deformation textures*.

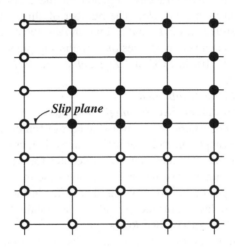

FIGURE S.2 Open circles show atomic positions before slip deformation; solid circles correspond to the positions after it. For further details, see *deformation twinning*, Figure D.4.

slip band Wavy *slip trace* that can be observed inside a *grain* on a polished surface of a *plastically deformed* specimen. Slip band is thicker than *slip line* and results from *multiple cross-slip*.

slip direction *Lattice direction* coinciding with the *Burgers vector* of *perfect dislocations* in the lattice concerned. Usually, it is a *close-packed direction*, e.g., $\langle 110 \rangle$ in *FCC*, $\langle 111 \rangle$ in *BCC*, and $\langle 11\bar{2}0 \rangle$ in *HCP* structures.

slip line *Slip trace* inside a *grain* that can be observed with the aid of an *optical microscope* on a polished surface of a *plastically deformed* sample. It is associated with a small step produced on the sample surface by *slip* over a single *slip plane*. In materials with low *stacking-fault energy*, slip lines are straight because the *dislocations* are *splitted* and cannot *cross-slip*. Slip lines are wavy if cross-slip takes place during deformation, e.g., in materials with high stacking-fault energy (see *pencil glide*) or in materials with low stacking-fault energy under the influence of *thermal activation* (see *Shockley dislocation*).

slip plane *Lattice plane* over which *dislocations* can *glide*. Usually, but not necessarily, it is the *close-packed plane*. In metals with *FCC lattice*, slip planes are {111} and in *BCC* metals, {110}, {112}, and {123}; whereas in *HCP* metals, they are *basal*, *prism*, or *pyramidal planes*, depending on the *axial ratio*. In *ionic crystals*, slip can result in the occurrence of anion-anion (or cation-cation) pairs, which increases the electrostatic energy. Because of this, slip planes, e.g., in *NaCl structure*, are not the close-packed planes {100}, but {110} planes.

slip system Combination of a *slip plane* with a coplanar *slip direction*. The number of possible slip systems is the largest in *BCC structure* (up to 48, depending on temperature that affects the choice of active slip planes), and much smaller in *FCC* and *HCP* structures. The number and type of slip systems in HCP structure depends on the *axial ratio* (see *basal*, *prismatic*, and *pyramidal slip*). The number of slip systems in *ionic* and *covalent crystals* is always smaller than in *metallic crystals*. A family of slip systems is denoted by indices of both the slip plane and slip direction as, e.g., {111}⟨110⟩ in FCC and *ZnS cubic structures* or {110}⟨100⟩ in *CsCl structure*.

slip trace See *slip line* and *slip band*.

small-angle grain boundary See *low-angle boundary*.

smectic crystal *Liquid crystal* whose rod-shaped molecules are arranged with their long axes approximately parallel to each other, as in *nematic crystals*. However, in contrast to the latter, the molecules form layers, the thickness of the layer being approximately the same as the molecule length. There are various types of smectic crystals that differ in the molecule arrangement inside the layer. Smectic crystals are also called S crystals.

Snoek–Köster peak/relaxation See *Köster peak*.

Snoek peak/relaxation *Internal friction* peak observed in materials with *BCC structure* due to reorientation of *interstitial solute* atoms producing tetragonal *lattice distortions*. Measurements of Snoek relaxation are used for determining the *diffusion coefficient* and the *solubility limit* of interstitial solutes, as well as for studying the *strain-aging phenomenon*.

soaking *Heat treatment* of a *steel* semi-product before *hot deformation*, with the aim of dissolving *second-phase* inclusions and obtaining chemically homogeneous *austenite*. Soaking temperature lies within an *austenitic range*.

softening anneal *Heat treatment* of *plastically deformed* material resulting in a *hardness* decrease due to the *recovery* or *primary recrystallization*.

solid Characterized by the ability to keep its shape over an indefinitely long period. The *atomic structure* of solid bodies can be either *crystalline* or *amorphous*.

solidification Occurrence of a *crystalline phase* from a liquid one. It proceeds at a constant temperature when the number of *degrees of freedom*, *F*, equals zero and in a temperature range if $F > 0$ (see *Gibbs' phase rule*).

At cooling rates exceeding *critical*, an *amorphous phase* occurs, and the *phase transition* is called *vitrification*.

solidification point/temperature Temperature at which *solidification* starts upon cooling. It corresponds to the *liquidus* in a *phase diagram*. An increased cooling rate leads to the melt *undercooling*, and solidification can start well below the liquidus (see *nucleation*). If cooling rate exceeds *critical*, a *glassy phase* can form at the temperature known as *glass transition temperature*.

solid solubility Description of the maximum *solute concentration* in an equilibrium *solid phase*. Solid solubility can be limited when the maximum solute concentration is lower than 100%, or can be compete when it equals 100% (see *Hume-Rothery rules*). *Solid solutions* with a limited solubility can be *substitutional* or *interstitial* or both, whereas those with complete solubility are always substitutional. *Intermediate phases* always have a limited solid solubility.

solid solution *Phase* whose *concentration* range includes at least one pure *component*. In terms of *crystal structure*, solid solutions can be *substitutional* or *interstitial*, as well as *random* (i.e., *disordered*) or *ordered*. Multicomponent solid solutions can be simultaneously substitutional and interstitial: e.g., chromium dissolves in a Fe–Cr–C solution substitutionally whereas carbon dissolves interstitially. In terms of the maximum *solute* concentration, there can be solid solutions with different *solubility limits*.

solid solution strengthening/hardening Increase in strength due to *static lattice distortions* induced by *solute* atoms. These distortions inhibit the dislocation *glide* motion and increase the *yield stress*. The effect is proportional to the solute *concentration*, c, or to $c^{1/2}$ if the interaction between the *dislocations* and solutes is strong enough, as in the case of *interstitial solutes*.

solid-state sintering Sintering of particulate compacts without the intentional addition of low-melting *dopants;* however, low-melting phases may be present due to *impurities* as, e.g., *silica* in Al_2O_3 (this results in the appearance of a thin liquid layer at *grain boundaries* during *firing* and in a *glassy phase* after it). In the course of solid-state sintering, the *microstructure* evolves as follows. At an intermediate stage, the pores are arranged at the grain boundaries. On further sintering, small pores remaining at the boundaries "dissolve" via the *vacancy* migration along grain boundaries toward the larger pores or the free surface (see *Gibbs–Thomson equation*). Simultaneously, *normal grain growth* develops as the *drag force* exerted by the pores vanishes gradually (see *particle drag*). Normal grain growth can be inhibited by *solute drag*, due to *grain-boundary segregation* reducing the *grain-boundary energy* and *mobility* (as, e.g., MgO in Al_2O_3). In some cases, *abnormal grain growth* commences during solid-state sintering. A decreased drag force resulting from an accelerated "dissolution" of pores at the boundaries with a liquid phase layer can trigger abnormal grain growth. Due to an increased grain size in the matrix, the abnormal grain growth is usually incomplete, which results a *duplex grain size* after firing. In the absence of abnormal grain growth, microstructure is relatively *fine-grained* and *homogeneous*. The microstructure of multiphase sintered products is always fine-grained due to inhibition of abnormal grain growth.

solidus Locus of *melting temperatures* or that of chemical *compositions* of *solid phases* at equilibrium with a liquid phase. In *binary phase diagrams*, the locus is a line, and in *ternary* diagrams, it is a surface.

solubility limit Maximum amount of *solute* that can be dissolved in an equilibrium *solid solution* under definite temperature and pressure. Solubility limit usually diminishes with a decrease in temperature, according to the *solvus* line (surface) in a *phase diagram*. See *solid solubility*.

solute *Component* dissolved in a *solid phase*.

solute diffusion Diffusion of *solute* atoms whose rate depends on their type: diffusion of *interstitials* proceeds faster than that of *substitutional* atoms. See *interstitial diffusion* and *vacancy mechanism*.

solute drag See *impurity drag*.

solution treatment *Heat treatment* consisting of: heating an *alloy* to temperatures corresponding to a single-phase *solid solution* field in a *phase diagram*; holding at a fixed temperature until all *second phases* dissolve; and cooling at a *critical rate* to retain a *supersaturated solid solution*. This treatment is usually the first stage of *aging treatment*; the second stage is *precipitation treatment*.

solvent Base of a *solid solution*.

solvus Locus of chemical *compositions* of a *solid solution* at equilibrium with another *solid phase*. In *binary phase diagrams*, the locus is a line, and in *ternary* diagrams, it is a surface.

sorbite Obsolete name of *eutectoid* in Fe–Fe₃C *alloys* that forms at temperatures in the middle of the *pearlitic range*. Its current name is *fine pearlite*. It is characterized by a smaller *interlamellar spacing* than *pearlite*, due to a decreased diffusion rate of iron and metallic *alloying elements* at decreased temperatures. It is named after British scientist H. C. Sorby.

S-orientation One of the main *texture components*, {123}⟨634⟩, observed in *cold-rolled FCC* metallic materials with increased *stacking-fault energy*.

sorption See *adsorption*.

space group Set of *symmetry elements* determining both the macroscopic and microscopic symmetry of *point lattices*. There are 230 space groups derived from 32 *symmetry classes*. Realization of all the symmetry elements of a symmetry class brings a crystal face into its initial position. Realization of all the symmetry elements of a space group can bring the crystal face into a different position, but it is crystallographically identical to the initial one.

special carbide See *alloy carbide*.

special grain boundary Grain boundary whose *disorientation* is characterized by a low Σ value (usually, it is assumed that Σ ≤ 25) and whose plane coincides with a *close-packed plane* of the corresponding *CSL*, as, e.g., a *coherent twin boundary*. The *atomic structure* of special boundaries is quite ordered and can be described by a CSL model. In high-purity materials, they are characterized by lower energy and mobility than *general grain boundaries*. At slightly increased impurity content, the mobility of special boundaries is higher than the mobility of general boundaries, which is caused by smaller *segregation* at the special boundaries. At an increased impurity content corresponding to commercial purity, special boundaries — except coherent twin boundaries — and general boundaries are characterized by nearly identical properties. An increased resistance to both *grain-boundary sliding* and grain boundary cavitation (see *creep cavitation*) is specific for special boundaries. Special boundary is sometimes called CSL-boundary.

specific [interface] area Mean total area of *interfaces* in a unit volume. All the *microstructure* changes not connected with *phase transitions* are accompanied by a decrease of specific interface area resulting in a decrease of the total energy of interfaces. See, e.g., *grain growth*, *particle coarsening*, *spheroidization*, etc.

specific volume Quantity reciprocal to the material density.

sphalerite [structure] type See *zinc blende [structure] type*.

sphere of reflection See *Ewald sphere*.

spherical aberration Lens defect caused by a difference in the focal lengths for rays that pass along the lens axis and through its periphery. Because of this defect, a point in an object spreads out into a circle in its image.

spheroidal graphite See *nodular graphite*.

spheroidite See *spheroidized pearlite*.

spheroidization Alteration of the shape of *incoherent precipitates* from lamellar or rod-like to *equiaxed*. Spheroidization proceeds at increased temperatures due to a different solubility in the *solid solution* surrounding particle areas with various surface curvature. See *Ostwald ripening* and *Gibbs–Thomson equation*.

spheroidized pearlite Pearlite whose *cementite* lamellae are *spheroidized* into globular particles under the influence of *heat treatment* comprising either a long holding stage slightly below A_1 or alternate temperature changes around A_1.

spherulite Spherical *polycrystalline* aggregate of fiber *crystallites* aligned along the aggregate radii. It can be observed in partially crystallized *glasses*.

spinel Mineral of *composition* $MgAl_2O_4$.

spinel ferrite *Ferrimagnetic* oxide of *stoichiometry* MFe_2O_4, where M is a transition metal or Li, Mg, Al, Ga, Zn, or Cu. Spinel ferrites have a partially inversed *spinel structure*.

spinel [structure] type *Cubic crystal structure* of several oxides of *stoichiometry* AB_2O_4; it is identical to that of "normal" *spinel,* $MgAl_2O_4$, in which A-cations are divalent and B-cations are trivalent. In the normal spinel, the O^{2-} ions form an *FCC sublattice*, the B^{3+} ions form their own sublattice, occupying half of the *octahedral voids* in the first sublattice, and the A^{2+} ions form the third sublattice, arranging in 1/8 of the *tetrahedral voids* of the first one. In the "inverse" spinel, half of the B^{3+} ions change their positions with the A^{2+} ions in the normal spinel.

spinodal In a *phase diagram*, locus corresponding to the inflection points on the curve of the *free energy of solid solution, G,* vs. its *composition, c,* at different temperatures. Under the locus the second derivative $\partial^2 G/\partial c^2$ is negative and on the locus it equals zero. Spinodal in a *binary* phase diagram is shown in Figure S.3.

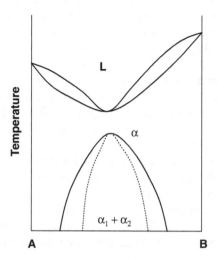

FIGURE S.3 Binary diagram with spinal decomposition. Dashed line shows a spinodal. For further details, see *binodal*, Figure B.2.

spinodal decomposition *Phase transition* on cooling in *crystalline solids* that results in decomposition of a *solid solution* into *metastable isomorphous* solid solutions of different chemistry. The *compositions* of the latter correspond to a *spinodal* in the *phase diagram* concerned (see Figure S.3). In spinodal decomposition, there is no energetic barrier to *nucleation,* and the decomposition proceeds through the occurrence and evolution of periodic compositional waves developing by *uphill diffusion. Microstructure* after spinodal decomposition consists of exceptionally fine *precipitates* (of 5–10 nm size, depending on the coefficient of *chemical diffusion*), homogeneously distributed in the *matrix.* This type of microstructure is called *modulated structure.*

spontaneous [Process] producing a reduction of *free energy.*

sputtering Atomization of a surface layer through the bombardment of primarily positive ions. It is used for *ion etching* and for a layer-by-layer chemical analysis (see, e.g., *SIMS*).

stabilized ZrO₂ Zirconia whose *polymorphic transformation* temperature is reduced to at least room temperature by certain *dopants*. There are three *polymorphic modifications* of pure zirconia with *monoclinic, tetragonal,* and *cubic lattices,* the *equilibrium temperatures, T_0,* being ~1240 and 2370°C for the transitions monoclinic form ↔ tetragonal form and tetragonal ↔ cubic, respectively. Some dopants, upon dissolving in ZrO₂, expand the field of *solid solutions* on the base of the tetragonal and cubic zirconia and shift T_0 to lower temperatures. Since the transformations tetragonal → monoclinic or cubic → tetragonal evolve as *martensitic*, the *martensite start temperature*, M_s, is much lower than T_0, and the corresponding high-temperature form may be preserved to the room temperature or below. For instance, 9 mol% CeO_2 decreases M_s down to −60°C (the material is known as Ce-stabilized ZrO_2), whereas *doping* with 2 mol% Y_2O_3 reduces it to temperatures below −196°C (the material is called Y-stabilized ZrO_2). The M_s temperature can also be decreased through *grain size* reduction in tetragonal zirconia. See *zirconia-toughened alumina*.

stable phase See *equilibrium phase*.

stacking fault *Planar crystal defect* characterized by an erroneous sequence of the *close-packed planes*, which is responsible for its increased energy. Inside *grains*, stacking faults are bordered by *partial dislocations*. See *intrinsic* and *extrinsic stacking faults*.

stacking-fault energy (SFE) Excess *free energy* of stacking fault resulting from distortion of the stacking sequence of *close-packed planes*. In the ratio $\gamma_i/Gb > 10^{-2}$, SFE is considered high, and at $\gamma_i/Gb < 10^{-2}$, it is assumed low (here, γ_i is the energy of *intrinsic stacking fault*, G is the *shear modulus*, and b is the *Burgers vector* of the *perfect dislocation*). See also *extrinsic stacking fault*.

staining Technique used for revealing *microstructure*. It consists of obtaining a transparent film on the polished surface of a *metallographic sample*. The light reflected from the upper and lower interfaces of the film produces interference colors that depend on the local film thickness and thereby reveal *grains* of *different phases* or various orientations as, e.g., after anodic oxidation of aluminum samples.

stair-rod dislocation See *Lomer–Cottrell barrier*.

standard (*hkl*) projection *Stereographic projection* displayed in such a way that the (*hkl*) *pole* is at its center.

standard triangle Part of the *standard (001) projection* for crystals of *cubic systems*; it has the shape of a curvilinear triangle, with vertices 0 0 1, 1 1 0, and 1 1 1. In *hexagonal* crystals, it is a triangular part of the standard (0001) projection, with vertices 0 0 0 1, 1 1 $\bar{2}$ 0, and 1 0 $\bar{1}$ 0. In such triangles, *poles* obtained by equivalent macroscopic *symmetry operations* are lacking. Standard triangle is also referred to as unit stereographic triangle.

static lattice distortion Local *short-range* change of *interatomic spacing* in *solid solutions* caused by a difference in the *atomic radii* of *substitutional solute* and *solvent*, or by a discrepancy between the radius of an *interstitial* solute and the radius of a *lattice void*. These changes induce strain fields leading to *solid solution strengthening*.

static recovery See *recovery*.

static recrystallization See *primary recrystallization*.

steady-state creep Stage of the creep process in which it evolves at a constant *strain rate*. See *power-law creep*, *Coble creep*, *Nabarro–Herring creep*, and *Harper–Dorn creep*. Steady-state creep is also referred to as *secondary creep*.

steel   Fe-based *alloy* whose *phase transitions* do not include an *eutectic reaction*.

steel martensite   Along with the features common for *martensite* in any *alloy system*, martensite in steels is characterized by the following special features: it occurs upon cooling *austenite*; if austenite is heated to the temperatures above A_3 or A_{cm}, the cooling rate should exceed *critical*. To receive martensite from *undercooled* austenite, a lower cooling rate is necessary (see, e.g., *martempering*). It is a *supersaturated solid solution* of carbon and *alloying elements* in *α-Fe* and has a *body-centered tetragonal lattice* with an *axial ratio* up to ~1.09, depending on the carbon content. Its crystallites are characterized by the *habit planes* $\{111\}_A$, $\{225\}_A$, $\{259\}_A$, or $\{3\ 10\ 15\}_A$ (see *Kurdjumov–Sachs, Greninger–Troiano*, and *Nishiyama orientation relationships*). It is characterized by a significantly greater *specific volume* in comparison to austenite (by ~3%). Steel martensite (except low-carbon steels and quenched *maraging* steels) is characterized by high *hardness* and brittleness as well as by low plasticity due to an increased carbon content in martensite. The *flow stress* is proportional to $\sim c^{1/2}$, where c is the carbon atomic fraction. This *strengthening* effect can result from: *static distortions* of the martensite lattice; an increased *dislocation density* in martensite; a highly specific area of *interfaces* inside martensite crystals, first of all, the boundaries of the *transformation twins*; carbon segregations at the dislocations; or the *precipitation* of small *carbides* from as-quenched martensite at increased M_f.

stereographic net   *Wulff* or *polar nets* used in the analysis of *crystallographic textures* and *single-crystal orientation*.

stereographic projection   Two-dimensional graphic presentation used for determining the angles between the *lattice planes* and *directions* in a *crystal*, as well as the crystal orientation with respect to an external coordinate system. Stereographic projection is obtained by locating the crystal at the center of the *reference* (orientation) *sphere* and displaying the *poles* of its lattice planes on the sphere and by projecting the *poles* from the south pole of the reference sphere upon the equatorial plane perpendicular to the north–south diameter. Measurements on the projection are fulfilled using the *Wulff net*.

stereology   See *quantitative metallography*.

stoichiometry/stoichiometric [composition]   Composition of *intermediate phases* or compounds that can be described by the formula A_mB_n, where A and B are *components*, and m and n are small integers, e.g., AB, A_2B, A_2B_3, etc. In *ionic crystals*, stoichiometry is determined by charges of cations and anions and maintains electrical neutrality. In *intermetallic phases* in which the valence rule is not fulfilled, stoichiometry is also observed. Deviations from stoichiometry can take place in the limits of a single-phase field in the *phase diagram*; they are referred to as nonstoichiometry. In *Hume–Rothery phases*, nonstoichiometry is associated with

the presence of *antistructural atoms* or *structural vacancies*. In ionic
compounds, it results from the appearance of *antisite defects* in the cation
sublattice and structural vacancies in the anion sublattice, or vice versa.
In some ionic compounds, e.g., $Fe_{1-x}O$ or TiO_{2-x}, nonstoichiometry may
result from a varying valence of cations.

stored energy In materials science, deformation energy absorbed by a *plastically
deformed* body and associated primarily with a *dislocation structure* forming in the course of deformation. It is usually 5–10% of overall deformation
work. The magnitude of stored energy is orientation-dependent, i.e., it is
different in various *components* of *deformation texture*. Stored energy
affects the development of *recovery* and *primary recrystallization*, whereas
its orientation dependence influences the *microstructure homogeneity* after
recrystallization annealing, as well as the *recrystallization texture*.

strain Relative change in the linear sample size at *tension* (*tensile strain*), in the
shape at *shear* (*shear strain*), or in the volume at *hydrostatic pressure*
(*dilation*).

strain aging Holding a temper-rolled material at temperatures as low as ambient,
which restores the *sharp yield point*. This effect can be explained by the
formation of *Cottrell atmospheres* at fresh (mobile) *dislocations* generated
by the preceding *temper rolling*.

strained-layer epitaxy Formation of a *heteroepitaxial film* whose *lattice misfit*
with the substrate is compensated by *coherency strains* only. Measures
preventing the penetration of *dislocations* from the substrate help to reach
the maximum thickness of the layer.

strain hardening Increase in the *true stress*, *s*, with *true strain*, γ, and *strain
rate*, $\dot{\gamma}$:

$$s = k\gamma^n\dot{\gamma}^m$$

where *k* is a coefficient, and *n* is the strain-hardening exponent, $n =$
0.1–0.5. The exponent *m* characterizes the strain-rate sensitivity; an
increase in the deformation temperature decreases *m*. Strain hardening is
a result of an increase in *dislocation density*, as well as of the *cell structure*
formation. An increase in the dislocation density, ρ_d, leads to an increase
of the *flow stress*, σ, by

$$\Delta\sigma = \beta G b \rho_d^{1/2}$$

where β is a constant, *G* is the *shear modulus*, and *b* is the *Burgers vector*.
The effect of cell structure is described by an equation similar to the *Hall–
Petch equation*. Strain hardening is also referred to as work hardening.

strain-hardening exponent See *strain hardening*.

strain-induced grain boundary migration (SIBM) Displacement of a *grain
boundary*, observed upon heating a slightly deformed material, under the
influence of the difference in the *dislocation density* or in the mean

subgrain size in the adjacent grains. The boundary, in this case, can move away from its center of curvature (compare with *capillary driving force*).

strain-induced martensite Martensite occurring below M_d and above M_s under the influence of *plastic deformation*. The occurrence of strain-induced martensite is responsible for *TRIP* effect.

strain rate Change in *true strain* per unit time.

strain rate sensitivity See *strain hardening*.

strain ratio See *r-value*.

Stranski–Krastanov growth mode Growth mode of *heteroepitaxial films*, intermediate between those by *Frank–van der Merve* and *Vollmer–Weber*, i.e., the film nucleation evolves layer by layer, and its subsequent growth evolves by the formation and growth of islands. Stranski–Krastanov growth mode is usually observed if the film and substrate have identical *bond* types.

stress In materials science, a quantity equal to the magnitude of an external force per unit surface of the flat area it acts upon. Depending upon the force direction in respect to the area, there can be *normal* and *shear stresses*. Stress units are MN/m^2 (MPa). See also *nominal stress* and *true stress*, as well as *tensile stress*.

stress-assisted martensite Martensite occurring below M_s^σ but above M_s under the influence of *elastic deformation* that assists in transforming martensite *embryos* into martensite *nuclei*. The occurrence and growth of stress-assisted martensite can be accompanied by a noticeable deformation (see *pseudoplasticity*). Stress-assisted martensite is also called stress-induced martensite.

stress–deformation diagram *Nominal stress* versus *nominal strain* dependence obtained in *tension* (compression) testing.

stress-induced martensite See *stress-assisted martensite*.

stress relaxation Decrease of *elastic deformation* with time at a constant applied load or at a constant absolute deformation. Stress relaxation is caused by *diffusion* that leads to the *dislocation* rearrangements and facilitates the *plastic deformation* (see, e.g., *dislocation creep*). As a result, stresses in the deformed body decrease.

stress-relief anneal *Annealing treatment* well below the *recrystallization temperature* resulting in a partial reduction of *macroscopic residual stresses* due to *stress relaxation*.

stress–strain diagram *True stress* vs. *true strain* dependence.

stretcher-strain marking Array of *Lüders bands* observed in *coarse-grained* low-carbon *steel* sheets subjected to cold drawing or cold stretching.

striation structure See *cellular substructure*.

structural disorder Defect *crystal structure* of an *intermediate phase*, e.g., an *electron compound* or a compound with *ionic bond* in which there are unoccupied sites in one of its *sublattices* (known as *structural vacancies*) or atoms at the sites in a "wrong" sublattice (known as antistructural atoms or antisite defects). Structural disorder helps to maintain a constant *electron concentration* in electron compounds and the charge balance in *ionic crystals.* For instance, an anion *vacancy* in the latter must be compensated by a cation vacancy (see *Schottky pair*) or by a cation *self-interstitial* (see *Frenkel pair*). Structural disorder is frequently accompanied by nonstoichiometry (see *stoichiometry*). It is also referred to as defect structure.

structural vacancy Unoccupied site in one of the *sublattices* of an *intermediate phase,* usually a *Hume–Rothery phase,* or a phase with *ionic bond.* See *structural disorder.*

structure Can refer to *phase composition,* but more frequently refers to *microstructure* or *substructure.*

structure factor/amplitude Quantity characterizing the intensity of an x-ray (electron) beam diffracted from a *crystal lattice* under the *Bragg conditions.* The shape and size of its *unit cell* are of no importance. If the *Bravais lattice* is *primitive,* the intensities of all the possible reflections are nonzero; in non-primitive Bravais lattices, some reflections can be of zero intensity. Combinations of the indices of *lattice planes* yielding zero intensity are known as *extinction rules.* Structure factor is taken into account in *x-ray structure analysis.*

structure-insensitive Determined primarily by the *phase composition* and affected by *microstructure* only slightly, as, e.g., saturation magnetization, coefficient of thermal expansion, etc.

structure-sensitive Affected by both the *microstructure* and *phase composition,* as, e.g., all mechanical properties, magnetic hysteresis, etc.

subboundary Boundary between *subgrains.* Since its *disorientation* angle is always small, subboundary is a *low-angle boundary.*

subcritical annealing In *hypoeutectoid steels,* an annealing below Ac_1 aiming at, e.g., *cementite spheroidizing.*

subgrain Part of a *grain* disoriented slightly with respect to the adjacent parts of the same grain.

subgrain boundary See *subboundary.*

subgrain coalescence Merging of neighboring subgrains. It could take place if the *subboundary* between subgrains becomes unstable, e.g., due to redistribution of *microscopic stresses* during *recovery.* A *grain boundary* bordering the coalescing subgrains can act as a sink for the *dislocations* stemming from the disappearing subboundary. Subgrain coalescence is supposedly a preliminary stage of the *recrystallization nucleus* formation.

subgrain structure *Substructure* formed by subgrains. It can occur on *solidification, recovery* (especially on *polygonization*), *continuous recrystallization,* or *phase transitions.*

sublattice Constituent of a *crystal lattice.* For instance, the *unit cell* centers of *BCC lattice* form a *primitive cubic* sublattice, whereas the vertex sites belong to the other primitive cubic sublattice.

substitutional atom Foreign atom replacing a *host* atom in the *crystal structure.* Substitutional atoms relate to *point defects.*

substitutional solid solution Solid solution wherein the *solute* atoms replace *host* atoms in its *crystal structure.* See also *Hume–Rothery rules.*

substructure *Structure* characterized by the type, arrangement, and density of *dislocations* and their agglomerations (known as dislocation structure, e.g., *cell structure, dislocation pile-ups*, etc.) or by the size, shape, and *disorientation* of *subgrains* (it is known as fine structure or subgrain structure).

supercooling See *undercooling*.

superdislocation *Perfect dislocation* in *ordered solid solutions* or *intermediate phases*. Its *Burgers vector* is equal to the *translation vector* of the *superlattice* concerned. Superdislocation can split into partials whose Burgers vectors are identical to that of *perfect dislocations* in the corresponding disordered *crystal structure*. The partials border an *antiphase boundary*, which occurs as a result of the splitting.

superheating Difference in $T - T_0$, where T_0 is the *equilibrium temperature* of the transformation concerned, and T is the actual temperature of the transformation start upon heating, $T > T_0$. Superheating in solid-state transformations can reach several tens of degrees, whereas in melting, it is usually negligible.

superlattice See *ordered solid solution*. The term has its origin in the fact that additional lines are observed in x-ray diffraction patterns of an ordered solid solution.

superplasticity Ability to deform *plastically* up to *nominal strains* $\geq 1000\%$ without noticeable *strain hardening*. In the course of superplastic deformation (SD), high *strain-rate sensitivity* ($0.2 \leq m \leq 0.8$) is observed. SD evolves at increased temperatures (0.5–$0.7\ T_m$) and low *strain rates* (10^{-4} to $10^{-3}\ \mathrm{s}^{-1}$) provided the grain size does not change significantly. The *dislocation density* inside the grains does not increase in the course of SD, and the grains remain *equiaxed*. Superplasticity is observed in *ultra fine-grained, dual-phase* materials or in fine-grained single-phase materials, with or without *dispersed second-phase particles*. SD is a result of equilibrium between *strain hardening* due to *dislocation multiplication* inside grains during plastic deformation, on the one hand, and softening due to the dislocation absorption by the *grain boundaries*, where dislocations *delocalize* and *annihilate*, on the other. This effect of grain boundaries can reveal itself in *grain boundary sliding*, in ceramics promoted by thin *intergranular* layers of *glassy phases*. Superplasticity can also result from *dynamic recovery* in single-phase materials.

supersaturation Excess *concentration* of a *solute* in a *solid solution* above the concentration at equilibrium, c_e. The degree of supersaturation can be expressed as c/c_e, $c - c_e$, or $(c - c_e)/c_e$, where c is solute concentration.

superstructure In materials science, the same as *ordered solid solution*. In microelectronics, a layered structure consisting of *single-crystalline epitaxial* films (thickness of several monoatomic layers) of different semiconductor substances is also called superstructure or, sometimes, heterostructure. It is characterized by highly perfect *crystal lattices* of its layers and *coherent interfaces* between the layers.

surface-energy driving force Additional driving force for *grain boundary* migration in films and tapes resulting from the contribution of decreased energy of the free surface:

$$\Delta g = 2\Delta\gamma_s/\delta$$

Here, $\Delta\gamma_s$ is the difference in the surface energy of two neighboring grains, and δ is the object thickness. This driving force can act in thin objects with a *columnar structure* whose thickness is low enough for the free surface area to be significant with respect to the total *interface* area (including grain boundaries) per unit volume.

surface tension See *interfacial energy.*

Suzuki atmosphere Increased or reduced *concentration* of *substitutional solutes* in *stacking faults* in comparison to the *lattice* with a perfect stacking sequence. Suzuki atmosphere can form in cases in which the interaction energy between the solute and the *solvent* atoms, and thus solubility, depends on the stacking sequence. In contrast to *Cottrell atmospheres*, Suzuki atmosphere is a result of a chemical interaction only.

symmetric boundary *Tilt* grain boundary whose plane divides the angle between identical planes in the lattices of the adjacent *grains* into two equal parts. Such a boundary coincides with the *close-packed plane* in the corresponding *CSL.*

symmetry axis One of the macroscopic *symmetry elements.* It is an axis passing through a point in a *point lattice* and bringing the lattice into self-coincidence by rotation about the axis. The angle of rotation can be 180, 120, 90, or 60° only (see *n-fold axis*).

symmetry class Set of *symmetry elements* determining the macroscopic symmetry of *point lattices.* Since the number of the possible symmetry elements exceeds the minimum set of the elements in each of the seven *crystal systems*, there is a total of 32 symmetry classes. Symmetry class is also called point group.

symmetry element *Symmetry axis*, *mirror plane*, inversion center, and rotation-inversion axis corresponding to the macroscopic *symmetry operations.* There are several microscopic symmetry elements, as, e.g., glide plane, translation, etc.

symmetry operation Macroscopic operation performed on a *point lattice* that brings it in self-coincidence. Macroscopic symmetry operations are rotation, reflection, inversion, and rotation-inversion.

system In thermodynamics, a system is an aggregate of bodies. A thermodynamic system can be open (if there is both the energy and mass exchange with other systems), closed (if there is no mass exchange with other systems), *adiabatic* (if there is thermal energy exchange with other systems), and isolated (if there is no exchange at all). In materials science, a system is a combination of various amounts of given *components*; it is referred to as an alloy system.

T

Taylor factor Quantity averaging the influence of various *grain* orientations on the *resolved shear stress*, τ_r, in a *polycrystal*:

$$\sigma = M\tau_r$$

(*M* is the Taylor factor, and σ is the *flow stress*). The averaging is fulfilled under the supposition that the deformations of the polycrystal and its grains are compatible. Reciprocal Taylor factor can be used for polycrystals instead of *Schmid factor,* whose magnitude is defined for a single grain only. In a nontextured polycrystal with *FCC structure*, reciprocal Taylor factor is 0.327.

temper carbon In *malleable irons*, *graphite* clusters varying in shape from flake aggregates to distorted nodules.

tempered martensite *Microconstituent* occurring in quenched *steels* upon the *tempering treatment* at low temperatures. Due to the precipitation of ε-*carbides*, the lattice of tempered martensite is characterized by a *tetragonality* corresponding to ~0.2 *wt%* carbon dissolved in the martensite. See *steel martensite*.

tempering of steel martensite Alterations in the *phase composition* under the influence of *tempering treatment*. They are the following. Up to ~200° C, as-quenched *martensite* decomposes into *tempered martensite* and ε- (or η-) *carbide* (in low- to medium-carbon steels) or χ-*carbide* (in high-carbon steels). Above ~300°C, *cementite* precipitates from the tempered martensite, whereas the latter becomes *ferrite* and the ε- and η- (χ-) carbides dissolve. In steels alloyed with *carbide-formers*, the *alloying elements* inhibit the carbon diffusion and displace all the previously mentioned phase transitions to higher temperatures. In addition, at temperatures ~600°C, the diffusion of the *substitutional* alloying elements becomes possible, which leads to the occurrence of *special carbides* accompanied by cementite dissolution. The phase transformations described are accompanied by the following microstructural changes in martensite and ferrite. Crystallites of tempered martensite retain the shape of as-quenched martensite. Ferrite grains, occurring from tempered martensite, do not change their elongated shape and *substructure* until *coars-*

ening and *spheroidization* of cementite precipitates starts, although *dislocation density* decreases and *subgrains* form inside the ferrite grains. Further heating leads to the subgrain growth and, eventually, to *continuous recrystallization*, which makes the ferrite grains look *equiaxed*. As for carbides, η- and ε-carbides form inside the martensite laths, the former appearing as rows of ~2 nm particles and the latter as thin *Widmannstätten* laths. At higher temperatures, while ε- and η-carbides dissolve, thin Widmannstätten plates of cementite occur inside the ferrite grains; they nucleate at the *interfaces* between ε- or η-carbide and ferrite. Simultaneously, a certain amount of cementite occurs at the ferrite grain boundaries. Heating above 300–400°C is accompanied by spheroidization and coarsening of cementite particles, especially those at the grain boundaries. Special carbides in alloy steels nucleate at the interfaces between cementite and ferrite and at the grain boundaries and subboundaries in ferrite, as well as on the dislocations inside the ferrite grains. Due to the low *diffusivity* of substitutional alloying elements, the precipitates of special carbides are smaller than cementite particles.

tempering of titanium martensite Changes in *phase composition* accompanying *decomposition* of titanium martensite upon heating. The changes depend on the martensite *structure*, the content and type of the *alloying elements*, and treatment temperature. For instance, in β-*isomorphous alloys* with a decreased *solute* content, α′-*martensite* transforms directly into α-*phase* of an equilibrium *composition* and fine precipitates of β-*phase* at the boundaries of the martensite laths, as well as inside them. In alloys with low M_s and high solute content, α″-*martensite* transforms into β-phase, which subsequently decomposes at increased temperatures (see *aging in Ti alloys*).

tempering [treatment] *Heat treatment* of *quenched alloys* aimed at obtaining a more appropriate combination of strength and toughness than after quenching. Tempering of *steels* consists of heating a quenched article to temperatures below Ac_1, holding at a chosen temperature, and subsequent cooling at a definite rate. This treatment is fulfilled immediately after *quenching* because *thermal* and *transformation stresses* may lead to the article's destruction. For tempering treatment of Ti-based alloys, see *aging treatment of Ti alloys*.

temper rolling *Plastic deformation* of *annealed steel* sheets by passing them through a rolling mill, where the sheet is slightly bent successively in two opposing directions. This operation is aimed at unlocking *dislocations* and suppressing the occurrence of *Lüders bands*. This effect is temporal because *strain aging* after temper rolling restores the *sharp yield point*.

tensile strain Relative length change, with respect to the gage length, measured along the sample axis in *tension* tests.

tensile stress See *tension*.

tension Loading scheme wherein two parallel and equal external forces are applied to a sample along its axis. An increase of the forces leads to the sample elongation and, simultaneously, to a decrease of its transverse size (see *Poisson's ratio*). The cross-section of the sample is assumed flat, and the normal *stresses* acting on the section are assumed to be distributed homogeneously. These stresses are often referred to as *tensile stresses*.

terminal solid solution Phase of a varying *composition* whose field in a *phase diagram* includes only one pure *component*. It is also referred to as primary solid solution.

ternary Consisting of three *components*.

tertiary cementite *Microconstituent* formed by cementite precipitating from *ferrite* in low-carbon *steels* upon slow cooling from temperatures lower than A_1. This is a result of a decrease of the carbon *solubility* in ferrite according to the PQ line in Fe–Fe$_3$C diagram. Tertiary cementite can be noticed at ferrite *grain boundaries* in steels containing <0.05 *wt%* C.

tertiary creep Creep stage characterized by a *strain rate* increasing with time. *Grain-boundary sliding* plays an important role at the creep stage, especially in *crystalline ceramics*, where it is promoted by *glassy phases* at the *grain boundaries*.

tertiary recrystallization *Abnormal grain growth* that evolves in thin sheets or films with *columnar microstructure* and is caused by *surface-energy driving force*. Since it was originally observed after preceding *secondary recrystallization*, it was called tertiary recrystallization.

tetragonal system *Crystal system* whose *unit cell* is characterized by the following *lattice parameters*: $a = b \neq c$, $\alpha = \beta = \gamma = 90°$.

tetragonality Ratio c/a in *tetragonal lattice*.

tetrahedral interstice See *tetrahedral void*.

tetrahedral site See *tetrahedral void*.

tetrahedral void In *crystal structures*, a *void* surrounded by four atoms that form the vertices of a tetrahedron. If rigid spheres of equal radii represent the atoms, the void radius is the maximum radius of an undistorted sphere located inside the void and touching the nearest atoms. In *FCC* and *HCP* *structures*, the number of tetrahedral voids is twice as large as the number of atoms. Tetrahedral void is also called tetrahedral site or tetrahedral interstice.

tetrakaidecahedron Polyhedron with eight hexagonal faces and six square faces obtained by a truncation of the regular octahedron apexes (see Figure T.1).

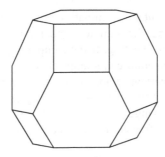

FIGURE T.1 Tetrakaidecahedron.

Tetrakaidecahedra of equal size can fill the space perfectly. Moreover, their faces can be bent in such a way that the angles between them in *triple junctions* and in quadruple points will correspond to the equilibrium of identical surface tensions. (This slightly distorted tetrakaidecahedron is known as Kelvin's tetrakaidecahedron.) It was supposed that the *equiaxed* grains in a single-*phase polycrystal* have a shape of the Kelvin's tetrakaidecahedron. However, this proposal was proved incorrect because the grains in the single-phase *microstructure* are always of noticeably different sizes.

texture See *crystallographic texture.*

texture analysis Experimental measurements of *crystallographic texture* and its subsequent representation by *pole figures*, *inverse pole figures*, or *orientation distribution functions*, followed by the identification of ideal orientations (*texture components*) and the estimation of their *intensity* and *scatter.*

texture component Orientation corresponding to a maximum in *pole figure* or *orientation distribution function*. The number of texture components is usually greater than one, although texture can be single-component as, e.g., a *cube texture*. It is also known as ideal orientation.

texture intensity Volume fraction of *grains* with a definite *ideal orientation* or of all the preferably oriented grains. The latter quantity characterizes an overall texture intensity.

texture scatter Deviation from an ideal orientation in a *texture component*. It can be measured as the width of a texture peak at a certain intensity level or as the standard deviation if the peak is represented by the Gauss function. Texture scatter is also referred to as orientation spread.

theoretical strength *Stress*, τ_{th}, necessary to shift one part of a perfect *crystal* with respect to the other part along a direction over the distance b:

$$\tau_{th} \cong Gb/(2\pi a)$$

where G is the *shear modulus*, b is the *interatomic spacing* in the shear direction, and a is the *interplanar spacing* for the shear plane concerned. This *shear* strength is minimum at the smallest b, i.e., for the *close-packed direction* and at the greatest a, i.e., for the *close-packed plane*. Compare with *Peierls stress.*

thermal analysis Experimental determination of the temperatures of *phase transitions* by measuring the sample temperature versus time upon slow, continuous cooling or heating. If the sample is chemically homogeneous and there are no temperature gradients, changes in the slope of the cooling (heating) curve at certain temperatures correspond approximately to the *equilibrium temperatures*; the corresponding temperatures are known as arrest points. The changes are connected with the latent heat evolution accompanying phase transitions upon cooling or with the heat absorption upon heating.

thermal etching Technique for revealing *microstructure* by heating an electropolished specimen in a nonoxidizing atmosphere. *Thermal grooves* occurring on the specimen surface delineate boundaries of *grains* and *incoherent twins*, as well as *phase boundaries*, and make the microstructure visible. Moreover, this technique gives the opportunity to trace the interface motion by observing the subsequent positions of thermal grooves. Thermal etching is also known as vacuum etching.

thermal groove Groove on a free surface along its intersection with a *grain* (or *phase*) *boundary*. Thermal grooves occur due to *thermal etching* and result from the tendency to equilibrate the energies of the free surfaces of two adjacent grains, γ_s, and that of the grain (phase) boundary, γ_{gb}. The dihedral angle at the grove bottom, ψ, can be found from the equation:

$$\gamma_{gb} = 2\gamma_s \cos (\psi/2)$$

on the condition, that γ_s for either grain is the same. For *high-angle boundaries, incoherent twin boundaries,* and phase boundaries in metals, $\gamma_{gb} \cong \gamma_s/3$ and thus $\psi \cong 160°$. Thermal grooves can inhibit grain boundary migration in thin tapes and films (see *groove drag*).

thermal hysteresis See *transformation hysteresis.*

thermally activated [*Spontaneous* process] evolved under the influence of a temperature elevation.

thermally hardened Strengthened by *heat treatment.*

thermal stability Concept including not only *thermodynamic stability* of certain *phases*, but also the *microstructure* stability.

thermal stresses *Macroscopic residual* stresses arising due to differences in thermal expansion in different parts of an article, e.g., in its surface layer and in the core.

thermal treatment See *heat treatment.*

thermodynamic equilibrium State of an isolated thermodynamic *system* characterized by lack of any temporal alterations and corresponding to the minimum *free energy* of the system. Compare with *metastable state.*

thermodynamic stability Absence of any *phase* changes in a *system* due to its *thermodynamic equilibrium.*

thermoelastic martensite Martensite *crystallite* changing its size simultaneously with temperature alterations between M_s and M_f. Such a behavior can be observed, provided martensite grows slowly upon cooling and there is no *stress relaxation* associated with its formation. These stresses preserve the atomic structure of the *interface* between martensite and the parent phase, which in turn makes possible a reversible motion of the interface synchronized with thermal cycles.

thermo-magnetic treatment Heat treatment comprising, as a last operation, a slow cooling in a magnetic field from temperatures above the *Curie point.* Such a treatment produces *magnetic texture.*

thermo-mechanical processing See *thermo-mechanical treatment.*

thermo-mechanical treatment Manufacturing procedure combining *plastic deforming* and *heat treating* with the aim of controlling *mechanical prop-*

erties through *microstructure* control. The latter is normally reduced to both *grain refining* in the *matrix* and a decrease of the *mean size* of *precipitates*, if present. See, e.g., *ausforming, isoforming, low-* and *high-temperature thermo-mechanical treatments.*

thickness fringes In *bright-field TEM* images, alternate bright and dark parallel bands. They occur due to an oscillating thickness dependence of the intensity of a diffracted electron beam. Thickness fringes are observed if there are *planar lattice defects* (*grain* and *twin boundaries, interfaces,* or *stacking faults*) inclined to the foil surface. Thickness fringes are arranged parallel to the intersection of the surface with a defect plane. They are also called wedge fringes.

thin foil Sample used for studying *microstructure,* chemical, and *phase compositions,* as well as *atomic structure* by *TEM, AEM, HVEM,* and *HRTEM.*

Thompson tetrahedron Geometric construction used for describing interactions of different *dislocations* in *FCC lattice.* The faces of the tetrahedron are {111} planes, and its edges are *Burgers vectors* of *perfect dislocations,* 1/2 ⟨110⟩.

Thomson–Freundlich equation See *Gibbs–Thomson equation.*

tie line Isotherm inside a two-*phase* field whose ends lie at intersections with the field boundaries. The ends give the *compositions* of the phases concerned. As shown in Figure T.2, intersections of a tie line in a (α + β)-phase field with the α/(α + β) and (α + β)/β *transus* (points *a* and *b*, respectively) yield the compositions of α- and β-phases at temperature, *T.* In *tertiary* diagrams, tie lines are to be taken on isothermal sections only. Tie lines can be used for determining the phase amounts in two-phase *alloys* by means of the *lever rule.* Tie line is also referred to as conode.

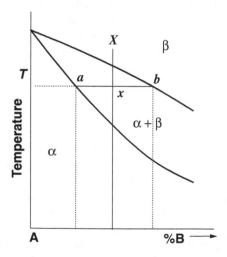

FIGURE T.2 Part of a binary phase diagram with the tie line *ab.*

tilt grain boundary Grain boundary whose *disorientation* axis is parallel to the boundary plane. In terms of the *grain boundary orientation*, there can be *symmetric* and *asymmetric* tilt boundaries.

time-temperature-transformation (TTT) diagram Presentation, in coordinates temperature–time, of the *kinetics* of *phase transformations* by lines corresponding to the onset and to the completion of the transformations. If the transformation is *athermal*, the corresponding lines are horizontal (see Figure T.3). TTT diagram is also referred to as isothermal transformation diagram.

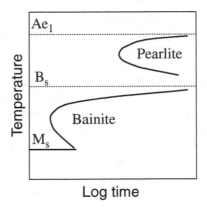

FIGURE T.3 TTT diagram for eutectoid alloy steel (scheme).

tinting See *staining*.

titanium martensite Product of the *martensitic transformation* $\beta \rightarrow \alpha'$ or $\beta \rightarrow \alpha''$. The phases α' and α'' are *supersaturated solid solutions* of *alloying* elements in α-*Ti* and differ in the *crystal system*: α'-martensite is *hexagonal*, whereas α''-martensite is *orthorhombic*. In addition, α''-martensite occurs at a greater *concentration* of β-*stabilizers* in β-phase. The *Burger orientation relationship*, as well as the *habit planes* $\{334\}_\beta$ or $\{441\}_\beta$ is observed between β- and α'-phases. Since titanium martensite is a *substitutional solid solution*, its formation is not accompanied by such a drastic strengthening, as in the case of *steel martensite*. Crystallites of titanium martensite are usually of a lath shape.

transcrystalline Observed or evolving inside *grains*. Intracrystalline, intragranular and transgranular are synonyms for transcrystalline.

transcrystallization zone See *columnar zone*.

transformation hysteresis Difference between the temperature of the start of a *phase transition* upon heating and that upon cooling (see *superheating* and *undercooling*). It is also referred to as thermal hysteresis.

transformation-induced plasticity (TRIP) Increase in both the ductility and *strain hardening* of *austenitic steels* caused by the formation of *strain-induced martensite*. An increase in ductility can be explained by the fact that *martensitic transformation* is geometrically analogous to plastic deformation (see *shear transformation*). Moreover, the stress relaxation accompanying the martensite formation prevents the initiation of cracks. An increase in strain hardening is connected with the fact that martensite has a high density of transformation-induced *lattice defects*, as well as with the *carbide precipitation* inside the martensite crystallites, which leads to *dislocation pinning*. See also *transformation toughening*.

transformation range Temperature range wherein a *phase transition* takes place. See, e.g., *pearlitic range, martensitic range*, etc.

transformation rate Rate of changes in the volume fraction of transformation products at a constant temperature.

transformation stresses *Internal stresses* arising as a result of a *specific volume* difference between a parent phase and the products of its transformation. Transformation stresses can be macroscopic if the transformation kinetics differ in various parts of a body.

transformation toughening An increase of the resistance to the crack growth due to the *phase transformation* triggered by the stress field at the crack tip. It is used in *crystalline ceramics*, e.g., in *zirconia-toughened alumina*. See also *transformation-induced plasticity*.

transformation twin Member of a series of twins causing an inhomogeneous *shear* inside *martensite crystallites*. Transformation twins accomodate shape distortions accompanying the martensite *nucleation* and growth, and also maintain the *habit plane* invariant. See *martensitic transformation*.

transformed β structure ($β_{tr}$) In air-cooled α + β Ti *alloys*, a *microconstituent* formed by *Widmannstätten* α-*phase* grains with a small amount of β-*phase* between them. Upon cooling from the temperatures above β/(α + β) *transus* the microstructure of these alloys consists of a thin network of α-phase grains delineating the prior *grain boundaries* of β-phase, whereas the rest of the field is occupied by $β_{tr}$. If the same alloy is air-cooled from temperatures below β/(α + β) transus, two microconstituents are observed, i.e., globular grains of α-phase not transformed upon cooling (they are called primary α) and *equiaxed* islands of $β_{tr}$ occupying the prior grains of β-phase. The volume fractions of the microconstituents depend on the heating temperature.

transgranular See *transcrystalline*.

transient creep First *creep* stage when the *strain rate* decreases with time. It is also called primary creep.

transient phase A *metastable phase* occurring at an intermediate stage of *aging*. The sequence of phase changes in this case can be described as follows: *supersaturated solid solution* → *supersaturated solution* + *GP zones (or a transient phase)* → *supersaturated solution* + a more stable transient phase → *saturated solution* + *stable second phase*. The solution *supersaturation* decreases with the evolution of its *decomposition* (see *rule of stages*). The *precipitates* of a transient phase are always *coherent* or *partially coherent*, whereas those of the stable phase are mostly *incoherent*. Transient phase is also referred to as transition phase.

transition band See *deformation band.*

transition phase See *transient phase.*

translation group See *Bravais lattice.*

translation vector In a *crystal structure,* a vector joining two neighboring positions of identical atoms (ions) chosen in such a way that the translation along the vector does not distort the crystal structure. For instance, in α-Fe (*BCC crystal lattice*), the translation vector is $1/2\ a\langle 111\rangle$, whereas in *CsCl structure* (with the same lattice), it equals $a\langle 100\rangle$.

transmission electron microscope (TEM) Electron microscope providing the image of a *substructure* and *electron diffraction pattern* with the aid of high-energy (100–300 keV) electrons transmitted through a thin sample. Two types of specimens are used in TEM: *replicas* and *thin foils.* The image of the *microstructure* is formed by an *absorption contrast* in the first case (with the *resolution limit* ~10 nm) and by a *diffraction contrast* in the second one (with the resolution down to the *atomic size*); the image in the second case can be either *bright-field* or *dark-field.* Both the *electron diffraction patterns* and *selected area diffraction patterns* can be obtained using TEM. This microscope is usually used for studying certain *structure* details as, e.g., small *precipitates, dislocation networks,* and other dislocation structures, including their *Burgers vectors, point defect* agglomerates, *stacking faults, grain boundaries, antiphase boundaries,* and *interfaces. SAD patterns* are used to find, e.g., the orientation of small *crystallites* or their *orientation relationship* with the *matrix,* etc.

transmission Laue method X-ray technique wherein a source and a flat film (screen), registering the diffracted radiation, are placed on the opposite sides of the flat specimen.

transus Boundary between a single- and a two-*phase* field in a *phase diagram.*

triclinic system *Crystal system* whose *unit cell* is characterized by the following *lattice parameters*: $a \neq b \neq c$, $\alpha \neq \beta \neq \gamma \neq 90°$.

tridymite Middle-temperature *polymorphic modification* of *silica* known as "high" tridymite. It has a *monoclinic crystal structure* and transforms, upon cooling at an increased rate, into a *metastable* "middle" tridymite at 160°C, and upon further cooling, into a metastable "low" tridymite at 105°C, both of the transformations being *displacive.*

trigonal system *Crystal system* whose *unit cell* is characterized by the following *lattice parameters*: $a = b = c$, $\alpha = \beta = \gamma \neq 90°$. It is also referred to as rhombohedral system.

triple junction In materials science, a point in a two-dimensional structure or a line in a three-dimensional structure where three *grain boundaries* meet. The *atomic structure* of triple junctions is more distorted than that of the grain boundaries. Since *grains* form and impinge at rather high temperatures, e.g., during *solidification, diffusional polymorphic transformation, recrystallization* etc., the boundaries in triple junctions usually meet at the dihedral angles corresponding to the equilibrium of their energies. If the energies are identical, all the dihedral angles are equal to 120°. Triple junctions are also referred to as triple points.

triple point With regards to *grain-boundary* networks, this is the same as *triple junction*. In single-component *phase diagrams*, it is the point of an invariant equilibrium of three *phases*, e.g., solid, liquid, and gaseous phases (see *Gibbs' phase rule*).

trostite *Pearlite* occurring in the lower end of *pearlitic range*. The term is obsolete.

true strain Magnitude of *tensile strain* calculated as

$$\gamma = \ln(l/l_0)$$

where l and l_0 are the current and initial lengths, respectively, of the sample subjected to *tension tests* or to rolling deformation. The true strain and the *nominal strain*, ε_{nom}, are connected by the formula:

$$\gamma = \ln(1 + \varepsilon_{nom})$$

and thus, $\gamma \cong \varepsilon_{nom}$ at small ε_{nom}. Because of this, true strain is taken into account at large deformations.

true stress Stress calculated from the *tension* test data as the external force divided by the current cross-sectional area of the sample. Magnitudes of the true and *nominal* stresses, s and σ_{nom}, respectively, for a particular sample differ noticeably at large deformations, ε_{nom}, only because

$$s = (1 + \varepsilon_{nom})\sigma_{nom}$$

twin Part of a *grain* or *single crystal* whose *lattice* orientation can be described as the mirror reflection of the lattice of the other part in a definite *lattice plane*; the latter is known as twinning plane. This reflection is not related to the macroscopic *symmetry operations* of the lattice concerned. Geometrically, twin orientation can also be obtained by a definite *shear* over a *twinning system* (see *deformation twinning*). Twins can occur during *plastic deformation* (*deformation twins*), *recrystallization, grain growth* (*annealing twins*), *phase transformation* (*transformation twins*), and even during *solidification* or *devirtification* (growth twins).

twinned crystal *Grain* or *single crystal* containing *twins*.

twinning system Specific combination of a *lattice plane* and a coplanar direction, a *shear* on which can produce a *twin* (see *deformation twinning*). In the *BCC structure*, twinning system is $\{112\}\langle110\rangle$; in *FCC*, it is $\{111\}\langle112\rangle$, and in *HCP*, it is $\{10\bar{1}2\}\langle1\bar{2}10\rangle$. Twinning system is unique in the sense that only a polar shear can result in a twin orientation. For instance, in BCC structure, the shear over (112) $[11\bar{1}]$ produces a twin orientation, whereas the shear over the same plane along the opposite direction, $[\bar{1}\bar{1}1]$, does not.

twist disclination *Disclination* producing a lattice rotation around an axis perpendicular to its line.

twist grain boundary Grain boundary whose *disorientation* axis is perpendicular to the boundary plane.

two-phase structure *Microstructure* consisting of two *microconstituents*, either one being single-*phase*, as in *dual-phase* or *duplex* structures.

two-way shape memory effect See *shape memory effect*.

ultra-fine grained Characterized by a *mean grain size* (or that of the *matrix phase*) smaller than ~1μm.

undercooling Difference $(T_0 - T)$ or $(T_0' - T)$, where T_0 is the *equilibrium temperature* of a *phase transformation*, T is the actual temperature of the transformation start, and T_0' is the temperature of equilibrium between the *metastable* parent *phase* and a metastable new one of the same *composition* (see Figure U.1). Undercooling is also referred to as supercooling.

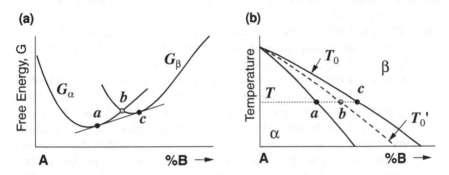

FIGURE U.1 (a) Free energies, G_α and G_β, of α and β solid solutions vs. concentration at temperature T. (b) Part of a binary phase diagram and corresponding equilibrium temperatures T_0 and $T_{0'}$. For further details, see *equilibrium temperature*, Figure E.1.

unit cell Parallelepiped that can be found by dividing space with three nonparallel sets of planes passing through the *lattice points* of a *crystal lattice*. Unit cell is chosen so that it has the same macroscopic *symmetry elements* as the *crystal system* of the lattice concerned. The unit cell shape is characterized by *fundamental translation vectors*, *a*, *b*, and *c*, directed along the cell edges. The lengths of these vectors, *a*, *b*, and *c*, respectively, are called unit cell parameters (constants), and the angles between the edges, known as axial angles, are denoted by α, β, and γ (see Figure U.2).

unit cell parameter See *lattice parameters*.

unit stereographic triangle See *standard triangle*.

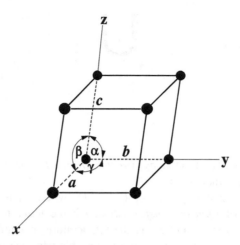

FIGURE U.2 Unit cell and its parameters.

unsaturated [solid] solution Solid solution whose *solute* content is below the *solubility limit*.

uphill diffusion Diffusion that increases the *concentration* gradient due to the *drift* flux decreasing the gradient of *chemical potential*. Uphill diffusion takes place, e.g., in the *spinodal decomposition*.

upper bainite *Microconstituent* occurring as a result of *bainitic transformation* of *undercooled austenite* in the upper part of *bainitic range*. It consists of packets of *ferrite* laths (the lath thickness ~0.2 μm), each lath being separated from the others by fine *cementite* particles. The laths nucleate at the austenite *grain boundaries*, and their *lattices* within the packet are disoriented only slightly. They stop growing into austenite before they impinge the grain boundaries due to the accumulation of *transformation stresses* in the austenite. See also *bainitic transformation*.

upper yield stress Maximum *tensile stress* attained before dropping down to the *lower yield stress*. See *sharp yield point*.

V

vacancy Unoccupied *lattice site* in a *host* lattice. Vacancies occur due to thermal *fluctuations*, as well as in the course of *plastic deformation, grain boundary migration, ion bombardment*, etc. At a certain *concentration*, vacancies decrease *free energy*, and their equilibrium concentration is:

$$c_v = n/N = \exp(-G_f/kT)$$

where n and N are the numbers of the vacancies and atom sites per unit volume, respectively, G_f is the *Gibbs free energy* of the vacancy formation, k is the *Boltzmann constant*, and T is the absolute temperature. Close to the *melting temperature*, the equilibrium concentration of vacancies in *crystalline* materials is 10^{-4}–10^{-6}. Vacancies can form associations with the other *point defects* (e.g., *divacancies* or pairs vacancy-*solute* atom); they disappear at *vacancy sinks* and occur at *vacancy sources*. Under certain conditions, e.g., after an abrupt cooling from high temperatures or after a heavy *plastic deformation* or because of *irradiation damage*, the vacancy concentration can greatly exceed the equilibrium concentration (see, e.g., *quench-in vacancies*). Nonequilibrium vacancies strongly increase the *diffusion* rate because they migrate to vacancy sinks. This is revealed in an increased rate of all *diffusion-controlled* processes at low temperatures (see, e.g., an *irradiation-induced creep*, the formation of *Guinier–Preston zones*, etc.) and in the occurrence of *grain-boundary segregations*. In semiconductors, vacancies are usually identical to *shallow impurities*. Vacancies should not be confused with *structural vacancies* in certain *intermediate phases* and in *ionic crystals*.

vacancy mechanism Diffusion mechanism of *substitutional solutes* and *solvent* atoms. Since a *vacancy* can jump from one lattice site into a neighboring one, a vacancy flux in one direction is equivalent to the atom transport in the opposite direction.

vacancy sink Site where vacancies disappear. *Edge* and *mixed dislocations*, as well as *planar crystal defects*, such as free surface, *grain boundary*, and *phase boundary*, can serve as vacancy sinks (see, e.g., *climb*).

vacancy source Element of *microstructure* generating vacancies. The main vacancy sources in *polycrystals* are *dislocations* and *grain boundaries*.

For instance, vacancies can be produced by moving *jogs* in the course of *plastic deformation* or by rapidly moving *grain boundaries*, especially during *primary recrystallization*. Vacancy sources can also serve as *vacancy sinks*.

vacuum etching See *thermal etching*.

valence band See *band structure*.

van der Waals bond Weak attraction between uncharged atoms (molecules) electrically polarized in the presence of each other. This bond type is weaker than *covalent, ionic,* or *metallic bonds*.

Vegard's law Empirical law, according to which, the *interatomic spacing* in a particular *substitutional solid solution* varies linearly with the atomic *solute concentration*.

vermicular graphite Short and thick *graphite crystallites* whose morphology is somehwhere between *nodular* and *flake*. It is also called quasi-flake, pseudo-nodular or compacted graphite.

vicinal plane Surface of a *single crystal* tilted about a certain direction with respect to a low-index lattice plane by an angle not greater than ~5°. As a result, vicinal plane is not smooth and is composed of relatively wide, atomically flat terraces connected by steps of several *atomic radii* height.

viscoelasticity See *anelasticity*.

vitreous phase See *glassy phase*.

vitrification Occurrence of a *glassy phase* upon cooling a liquid phase below a *glass ransition temperature*, i.e., at a cooling rate exceeding *critical*.

Vollmer–Weber growth mode Growth of a solid film via the formation of three-dimensional atomic clusters (known as islands) on the substrate and their sideways growth. In the course of their growth, the islands impinge and coalesce. The film grown according to this mode is called island film. It can be *single-* or *polycrystalline, epitaxial* or nonepitaxial, with *crystallographic texture* or without it, which depends on the film and the substrate material, substrate preparation, deposition technique and conditions, etc. This growth mode is usually observed if the interaction between the film and the substrate is weak, e.g., on depositing a semiconductor or metallic film on a dielectric substrate.

vol% Volume percentage. *Weight percentage* of a *phase* (or *microconstituent*) X, W_X, in a material consisting of two phases (microconstituents), X and Y, which can be calculated from its volume percentage, V_X, as follows:

$$W_X = 100/[1 + D_Y (100 - V_X)/(V_X D_X)]$$

(D_X and D_Y are the mass densities of X and Y phases or microconstituents, respectively).

volume diffusion See *bulk diffusion*.

W

Wagner–Lifshitz–Slyozov theory Theoretical description of the *Ostwald ripening*, according to which, the *mean size* of *equiaxed* precipitates, *d*, changes with time, *t*, at a constant temperature and a small volume fraction of *precipitates*, as follows:

$$d^3 - d_0^{\,3} = k(t - t_0)$$

The subscript 0 in the equation relates to the initial magnitudes of *d* and *t*, and the constant, *k*, is directly proportional to the energy of the *interface* between precipitates and the *matrix*, σ. In addition, *k* is proportional to the *diffusion coefficient* of *solutes* in the matrix, as well as to their *solubility limit*. The equation is usually referred to as $t^{1/3}$ dependence and is characteristic of *diffusion-controlled* growth. Since *coherent* and *partially coherent interfaces* are characterized by low σ, *coherent precipitates* coarsen slower than do *incoherent* precipitates. *Solute segregation* to the interface can also slow down the interface motion, which manifests itself in a change of the ripening kinetics from diffusion-controlled to *interface-controlled*, described by $t^{1/2}$ dependence.

warm deformation Procedure of *plastic deformation* at temperatures slightly below the *recrystallization temperature*.

warm worked Subjected to *warm deformation*.

Warren–Averbach method Technique for calculating the average size of coherently diffracting mosaic blocks (see *mosaic structure*) and the *microstrain* magnitude using the experimentally determined shape of *x-ray diffraction lines*, corrected for *instrumental line broadening*.

wavelength spectrum Distribution of radiation intensity vs. its wavelength.

wavelength-dispersive spectrometry (WDS) Technique for chemical analysis of single-*phase* materials using spectrometry of *fluorescent x-rays* emitted by the sample. Fluorescent radiation from the sample is directed on a *crystal monochromator* mounted on a *diffractometer* (instead of the specimen, as in the conventional *diffractometric method*) and rotated at a certain rate. A counter with a rotation rate two times greater than that of the monochromator analyzes the diffracted radiation. The recorded spectrum consists of lines with the same *Miller indices* corresponding to *characteristic x-rays* of different wavelengths emitted by different chemical elements in the sample.

weak-beam imaging *TEM* technique increasing the *resolution* of *lattice defects* in *dark-field images*. It is used primarily for imaging of *dislocations* and their agglomerations.

wedge disclination *Disclination* producing a lattice rotation around an axis parallel to its line.

wedge fringes See *thickness fringes*.

Weiss zone law See *zone*.

well-defined yield point See *sharp yield point*.

whisker Fibrous body with diameter ~1 μm and length of several mm. In *single-crystalline* whiskers, only one *screw dislocation* along the crystal axis can be present. Because of this, they are characterized by high strength, close to the *theoretical strength*, and by *elastic deformation* up to 4–5%.

white [cast] iron *Cast iron* whose typical *microconstituent* is a *ledeburite* and in which *graphite* is not observed. White irons are hard and brittle due to a large amount of *cementite*.

white-heart malleable [cast] iron Malleable iron with a *ferritic matrix*.

white radiation X-ray radiation with a *continuous wavelength* (*energy*) spectrum occurring due to the deceleration of electrons at the x-ray tube target. It is also referred to as bremsstrahlung or polychromatic radiation.

Widmannstätten ferrite Platelet-like grains of *proeutectoid ferrite* formed in *hypoeutectoid steels* with *Widmannstätten structure*. They are arranged regularly inside prior *austenite* grains, and a thin network of ferrite grains at the prior *grain boundaries* of austenite is also present. Widmannstätten ferrite grows via the migration of *coherent* or *partially coherent interfaces* of grains nucleating at the austenite grain boundaries. The *Kurdjumov–Sachs orientation relationship* with $\{111\}_A$ *habit plane* describes a relative orientation of austenite and ferrite. The boundaries of ferrite platelets, seeming flat under an *optical microscope*, contain, in fact, small ledges, and the platelets grow via a sideways motion of the ledges, i.e., their growth is *interface-controlled*.

Widmannstätten structure *Microstructure* occurring at an increased *undercooling* because of *diffusional phase transitions*, e.g., *polymorphic transformation* or *solid solution decomposition*. The grains of a new phase in the Widmannstätten structure grow within the grains of a parent phase, starting from a relatively thin network of *grain boundary allotriomorphs*. The new grains have a shape of thin plates or laths arranged parallel to definite planes of the parent phase, and their *lattice* is oriented in a definite way with respect to the parent phase lattice. The *aspect ratio* of Widmannstätten *precipitates* is commonly quite high, which can be explained, at least partially, by their *interface-controlled* growth.

work hardening See *strain hardening*.

wt% Weight percentage. *Atomic percentage* of component A, A_A, in a *binary system* A–B can be calculated from its weight percentage, W_A, by the formula:

$$A_A = 100/[1 + M_A (100 - W_A)/(W_A M_B)]$$

(M_A and M_B are the *atomic weights* of A and B, respectively). *Volume percentage* of a *phase* (or *microconstituent*) X, V_X, in a *heterogeneous* system can be calculated from its weight percentage, W_X, by the formula:

$$V_X = 100/[1 + \rho_Y (100 - W_X)/(W_X \rho_X)]$$

where ρ_X and ρ_Y are the mass densities of the X and Y phases (micro-constituents).

Wulff net Projection of longitude and latitude lines of a *reference sphere* on a plane containing the north–south axis of the sphere. See *stereographic projection*.

wurzite *Hexagonal polymorphic modification* of ZnS; it differs from the *cubic* form, sphalerite, by the sequence of atomic layers.

X

x-ray absorption spectrum Dependence of *mass absorption coefficient*, μ/ρ, on the x-ray wavelength, λ. The magnitude of μ/ρ at first increases continuously with λ, then decreases abruptly at the *absorption edge* (the edge is denoted by λ_K for the absorption of the K set, λ_L for the absorption of the L set, etc.), and again increases continuously. See *characteristic x-rays*.

x-ray diffraction (XRD) *Elastic* x-ray *scattering* accompanied by the occurrence of reflected x-ray beams whose directions are determined by the *Bragg law* or by the *Laue equations*. In materials science, XRD patterns are used for studying: *phase composition* in *heterogeneous systems*; *phase transformations*, e.g., during *aging*; *solid solution* type and the *solubility limit*; *short-range* and *long-range order* in *crystal lattice*; *lattice defects* and thermal vibrations in *single crystals*; *macroscopic* and *microscopic residual stresses*, etc.

x-ray diffraction line Peak in a *diffractogram* or an arc in a *Debye–Scherrer diffraction pattern*.

x-ray emission spectrum Wavelength spectrum of characteristic x-rays.

x-ray fluorescence Emission of *characteristic x-rays* under the influence of the primary x-ray (electron) radiation. The maximum fluorescence yield is observed if the wavelength of primary radiation is a little shorter than the *absorption edge* of the irradiated material (see *x-ray absorption spectrum*).

x-ray line intensity Maximum height of an *x-ray diffraction line* measured above the *background*.

x-ray line width Line width in *diffractogram*, measured in terms of 2θ at half the line height. It is also known as full width at half maximum.

x-ray microscopy See *x-ray topography*.

x-ray photoelectron spectroscopy (XPS) Technique for chemical analysis of *adsorption* layers. It utilizes photoelectrons emitted by the surface atoms under the influence of primary x-rays. XPS can detect the nature of the surface atoms as well as their chemical state.

x-ray scattering *Elastic* or *inelastic* scattering of x-rays in all directions by electrons in the material irradiated by x-rays.

x-ray spectroscopy Application of *Bragg's law* for determining the wavelength of x-rays diffracted on the specimen with a known *interplanar spacing*.

See *x-ray fluorescence*, *energy-dispersive*, and *wave-dispersive x-ray analyses*.

x-ray structure analysis Application of *x-ray diffraction* for determining *crystal structure*, i.e., *unit cell parameters*, *point group*, and *lattice basis*. In the analysis, the magnitudes of the *atomic scattering factor*, *absorption factor*, *Lorentz factor*, *polarization factor*, and *structural factor* are necessarily taken into account.

x-ray topography Techniques for investigating various *crystal defects* in nearly perfect crystals with a low defect density by studying the intensity distribution in x-ray *diffraction spots*. It is sometimes called x-ray microscopy.

Y

yield point elongation *Plastic deformation* observed after a *tensile stress* reduces from the *upper* to the *lower yield stress*. Yield point elongation commences with the appearance of *Lüders bands*, evolves without *strain hardening*, and completes when the bands spread over the entire specimen. At this stage, the material inside the Lüders bands is deformed plastically, whereas the material outside the bands is deformed *elastically*. Since the plastic strain is distributed nonhomogeneously, this deformation stage is also known as a discontinuous yielding. The magnitude of the yield point elongation increases with the test temperature. Yield point elongation is also known as the Lüders strain.

yield strength See *yield stress*.

yield stress *Tensile stress* corresponding to the onset of *plastic deformation*. In the absence of *sharp yield point*, yield stress is defined as a stress required for reaching a predetermined (usually 0.2%) *strain*. In the latter case it is called yield strength.

Young's modulus *Elastic modulus* measured at uniaxial *tension*. In *polycrystals* without *texture*, Young's modulus is *isotropic* and denoted by *E*. See *Hooke's law*.

Z

Zener drag *Drag force* for *grain boundary* migration exerted by *incoherent*, uniformly distributed particles of volume fraction *f* and *mean size d*:

$$\Delta g = a\gamma_{gb}f/d$$

where *a* is a coefficient, and γ_{gb} is the *grain-boundary energy*, the latter being assumed equal to the energy of the particle/grain *interface*.

Zener peak/relaxation Peak of *internal friction* observed in *substitutional solid solutions*. It results from the reorientation of either the *solute-solute* pairs in *dilute* solutions or the *host atom-vacancy* pairs in concentrated solutions. This relaxation type gives some information on the *diffusion mechanism* and *diffusion coefficients* at low temperatures, as well as on the changes in the *degree of long-range order*.

zinc blende [structure] type *Crystal structure* typical of many *ionic crystals*, as well as compounds with *covalent bonds*, e.g., GaAs, CdS, GaP, InSb, etc. It is identical to the *ZnS cubic structure*.

zirconia-toughened alumina (ZTA) *Composite* containing, apart from Al_2O_3, 5–10 vol% *dispersoid* of *stabilized tetragonal* ZrO_2. Its toughening is a result of *martensitic transformation* of the tetragonal zirconia into *monoclinic* form, initiated by the stress field at the crack tip in the alumina *matrix*. The transformation is accompanied by a significant volume increase and provides compressive stresses in the matrix that suppress the crack growth. As a result, the toughness of the composite exceeds that of pure alumina. See *transformation toughening*.

ZnS cubic structure *Crystal structure* of a *cubic polymorphic form* of ZnS; it is also known as zinc blende or sphalerite. It is composed of two *FCC sublattices*, one being occupied by the Zn atoms and the other by the S atoms. The sublattices are shifted by $1/4 \langle 111 \rangle$ with respect to one another (see Figure Z.1). Another way of describing the structure is the following: the S atoms form an FCC sublattice in which the Zn atoms form the other sublattice, whose sites occupy one half of the *tetrahedral voids* in the first one. A *coordination polyhedron* for both of the atom types in the structure is a tetrahedron, which implies that the *covalent crystals* with tetragonal

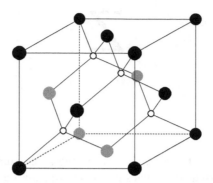

FIGURE Z.1 Unit cell of ZnS cubic structure. Solid spheres show S atoms forming one FCC sublattice, and open spheres show Zn atoms forming the other FCC sublattice (see text). Thick lines designate atomic bonds.

bond directionality may have the same structure. This is the case, e.g., in III-V semiconductor compounds.

zonal segregation See *macrosegregation*.

zone In crystallography, a family of *lattice planes* parallel to a definite *lattice direction* (it is known as zone axis). If a zone plane and zone axis have indices (*hkl*) and [*uvw*], respectively, then, according to the *Weiss zone law*:

$$hu + kv + lw = 0$$

zone annealing Procedure for producing large *single crystals* or *polycrystals* with coarse grains elongated in one direction. This is performed by *annealing ultra-fine grained* samples containing *dispersoids* in the temperature gradient moved along the specimen axis. *Coarse-grained* or *monocrystalline* structure forms as a result of *abnormal grain growth* controlled by *particle drag*.

zone axis In crystallography, a line of intersection of *lattice planes*.

Glossary

Glossary

English–German

60°-dislocation 60°-Versetzung

A

α-Al$_2$O$_3$ Aluminiumoxid
α-Fe α-Eisen
α isomorphous Ti system α-isomorphes Ti-System
α-phase [in Ti alloys] α-Phase [in Ti-Legierungen]
α'-martensite α'-Martensit
α"-martensite α"-Martensit
α-stabilizer alphastabilisierender Zusatz
α-Ti α-Titan
α Ti alloy α-Titanlegierung
(α + β) brass (α + β)-Messing
(α + β) Ti alloy (α + β)-Titanlegierung
A$_1$–Ae$_1$ temperature A$_1$–Ae$_1$-Temperatur
A$_2$–Ae$_2$ temperature A$_2$–Ae$_2$-Temperatur
A$_3$–Ae$_3$ temperature A$_3$–Ae$_3$-Temperatur
A$_4$–Ae$_4$ temperature A$_4$–Ae$_4$-Temperatur
A$_{cm}$–Ae$_{cm}$ temperature A$_{cm}$–Ae$_{cm}$-Temperatur
aberration Aberration; Abbildungsfehler
abnormal grain growth anormales, unstetiges, diskontinuierliches Kornwachstum; unstetige, anormale, diskontinuierliche Kornvergrößerung; Sekundärrekristallisation; Grobkornrekristallisation
abnormal pearlite anormaler Perlit
absorption Absorption
absorption coefficient Absorptionskoeffizient
absorption contrast Absorptionskontrast
absorption edge Absorptionskante
absorption factor Absorptionsfaktor
absorption spectrum Absorptionsspektrum
acceptor Akzeptor
accommodation strain Akkomodationsverformung, Kohärenzspannung
achromatic lens or objective Achromat
acicular nadelförmig
acicular ferrite nadelförmiger Ferrit
acicular martensite nadelförmiger, linsenförmiger Martensit; Plattenmartensit; Lattenmartensit
Ac temperature Ac-Temperatur

activation analysis Aktivierungsanalyse
activation energy Aktivierungsenergie
activation enthalpy Aktivierungsenthalpie
active slip system aktiviertes Gleitsystem, primäres Gleitsystem; Hauptgleitsystem
adatom adsorbiertes Atom; Adatom
adiabatic approximation adiabatische Näherung
adsorbate Adsorbat
adsorbent Adsorbent
adsorption Adsorption
after-effect Nachwirkung
age hardening Alterungshärtung; Aushärtung; Ausscheidungshärtung
aging Alterung; Aushärtung
aging [in Ti alloys] Aushärtung [in Ti-Legierungen]
aging treatment Auslagern; Altern; Aushärten
aging treatment of Ti alloys Aushärten von Ti-Legierungen
air-cooling Luftabkühlung
aliovalent solute or impurity aliovalenter Dotierstoff
allotropic change allotrope Umwandlung
allotropic form or modification allotrope Form, Modifikation
allotropy Allotropie
alloy Legierung
alloy carbide Sonderkarbid; Mischkarbid
alloying composition Vorlegierung
alloying element Legierungselement; Zusatzelement
alloy steel Sonderstahl
alloy system Legierungssystem
alpha brass Alpha-Messing
ambipolar diffusion ambipolare Diffusion
amorphous solid amorpher Festkörper; glasartige Substanz
amplitude contrast Amplitudenkontrast
analytical electron microscopy (AEM) analytische Durchstrahlungs-Elektronenmikroskopie
Andrade creep Andrade-Kriechen
anelasticity Anelastizität
anisotropic anisotrop
annealing or anneal Glühen; Weichglühen
annealing texture Glühtextur
annealing twin Rekristallisationszwilling; Glühzwilling
anomalous x-ray transmission anormale Röntgenstrahlungstransmission
antiferromagnetic Antiferromagnetikum
antiferromagnetic Curie point antiferromagnetische Curie-Temperatur; Néel-Temperatur
antiphase boundary Antiphasen-Grenze
antiphase domain Antiphasen-Domäne
antisite defect Antistrukturatom

antistructural atom Antistrukturatom
aperture diaphragm Aperturblende
apochromatic lens or objective Apochromat
arrest point Umwandlungstemperatur; Haltepunkt
Arrhenius equation Arrhenius-Beziehung
Ar temperature Ar-Temperatur
artifact Artefakt; Verfälschung
artificial aging Warmauslagern
asterism Asterismus
astigmatism Astigmatismus
asymmetric boundary unsymmetrische Korngrenze
athermal transformation athermische Umwandlung
atomic force microscope (AFM) Rasterkraft-Mikroskop
atomic mass Atommasse; Atomgewicht; Massenzahl
atomic packing factor Atompackungsfaktor
at% Atomkonzentration; At-%
atomic radius Atomradius
atomic scattering factor Atomformfaktor
atomic size Atomgröße
atomic structure Atomanordnung
atomic volume Atomvolumen
atomizing Schmelzverdüsung; Zerstäubung
atom probe field ion microscopy (APFIM) Atomsondenspektroskopie
Auger electron Auger-Elektron
Auger-electron spectroscopy (AES) Auger-Elektronenspektroskopie (AES)
ausforming Austenitformhärten; tieftemperatur-thermomechanische Behand-
    lung; Austenitverformen
austempering Bainitisierung; Zwischenstufenvergütung
austenite Austenit
austenite finish temperature (A_f) Temperatur vom Ende der Martensit-
    Aus-tenit-Umwandlung* (A_f-Temperatur)
austenite stabilization Austenitstabilisierung
austenite-stabilizer austenitstabilisierendes Element, Zusatz
austenite start temperature (A_s) Temperatur vom Anfang der Martensit-
    Austenit-Umwandlung* (A_f-Temperatur)
austenitic-ferritic steel austenitisch-ferritischer Stahl
austenitic-martensitic steel austenitisch-martensitischer Stahl
austenitic range Austenit-Temperaturbereich
austenitic steel austenitischer Stahl
austenitization Austenitisierung
autoelectronic emission Feldemission
autoradiography Autoradiographie
Avogadro number Avogadrosche Zahl; Avogadro-Konstante
Avrami equation Avramische Gleichung
axial angle Achsenwinkel
axial ratio Achsenverhältnis

B

β–Al$_2$O$_3$ β–Aluminiumoxid
β eutectoid Ti system β-eutektoides Ti-System
β-Fe β-Eisen
β isomorphous Ti system β-isomorphes Ti-System
β-phase [in Ti alloys] β-Phase [in Ti-Legierungen]
β-stabilizer betastabilisierender Zusatz/Element
β-Ti β-Titan
β Ti alloy β-Titanlegierung
background Untergrund
back-reflection Laue method Laue-Reflexionsmethode/-Rückstrahlverfahren
backscattered electron Rückstreuelektron
bainite Zwischenstufengefüge; Bainit
bainite start temperature (B$_s$) Temperatur vom Anfang der Bainitbildung* (B$_s$-
        Temperatur)
bainitic range Bainit-Temperaturbereich/-Temperaturintervall
bainitic transformation Zwischenstufenumwandlung; Bainitsche-Umwand-
        lung
bainitic steel bainitischer Stahl
bamboo structure Bambus-Gefüge
banded structure sekundäre Zeilengefüge; Fasergefüge
band gap Energielücke; Bandlücke
band structure Bänderstruktur
basal plane Basisebene
basal slip Basisgleitung
base Grundstoff; Basisstoff
base-centered lattice basiszentriertes Gitter; einseitig flächenzentriertes Gitter
based lattice basiszentriertes Gitter
Bauschinger effect Bauschinger-Effekt
bend contour Extinktionskontur; Biegekontur; Interferenzschliere
bicrystal Bikristall
bimetallic bimetallisch
bimodal bimodal
binary zweistoff
binodal Binodale
black-heart malleable [cast] iron schwarzer Temperguß; Schwarzkernguß
Bloch wall Bloch-Wand
blocky martensite lanzettförmiger, massiver Martensit; Blockmartensit
body-centered cubic (BCC) structure kubisch-raumzentriertes/krz Struktur
body-centered lattice raumzentriertes Gitter
Boltzmann constant Boltzmann-Konstante/-Faktor
bond energy Bindungsenergie
Bordoni peak or relaxation Bordoni-Dämpfungsmaximum/-Relaxation
Borrmann effect Borrmannsches Effekt
Bragg angle Braggscher Winkel; Glanzwinkel

Bragg [diffraction] condition Braggsche Reflexionsbedingung
Bragg reflection Braggsches Reflex
Bragg's law Bragg-Gleichung
brass Messing
Bravais lattice Bravais-Gitter
bremsstrahlung Bremsstrahlung; polychromatische Röntgenstrahlung
bright-field illumination Hellfeld-Beleuchtung
bright-field image Hellfeldbild; Hellfeldabbildung
bronze Bronze
Bs/Def orientation Messing-Lage
Bs/Rex orientation Messing-Rekristallisationslage
bulk diffusion Volumendiffusion
bulk modulus Kompressionsmodul
Burger orientation relationship Burger-Orientierungsbeziehung
Burgers circuit Burgers-Umlauf
Burgers vector Burgers-Vektor

C

χ-carbide χ-Carbid
CaF$_2$ structure CaF$_2$-Gitter; Flußspatgitter; Fluoritgitter
calorimetry kaloriemetrische Analyse; Kaloriemetrie
capillary driving force kapillare Triebkraft*
carbide Karbid; Carbid
carbide-former Karbidbildner
carbide network Karbidzellgefüge
carbide segregation Karbidseigerung
carbide stringers Zeilenkarbide
carbonitride Karbonitrid
cast iron Gußeisen
casting Abguß; Gußstück
C-curve C-Kurve
cell structure Zellgefüge
cellular microsegregation zellulare Mikroseigerung
cellular precipitation diskontinuierliche Ausscheidung
cellular substructure Zellengefüge
cementite Zementit
characteristic x-rays charakteristische Röntgenstrahlung
chemical diffusion chemische Diffusion; Interdiffusion
chemical etching chemische Ätzung
chemical inhomogeneity chemische Inhomogenität
chemical potential chemisches Potential
chemisorption chemische Adsorption
cholesteric crystal cholesterischer Kristall
chromatic aberration chromatische Aberration
cleavage plane Spaltebene

climb [Versetzungs] klettern
close-packed direction or row Gittergerade dichtester Besetzung; dicht-
    gepackte Atomreihe
close-packed plane Gitterebene maximaler Belegungsdichte
closing domain [magnetischer] Abschlußbereich; Abschlußbezirk
coagulation Koagulation
coalescence Vereinigung; Koaleszenz
coarse-grained grobkörnig
coarsening Vergröberung
coarse pearlite grober Perlit
Coble creep Coble-Kriechen
coherency strain Kohärenzspannung
coherency strain hardening Kohärenzspannungs-Verfestigung*
coherent interface kohärente Grenzfläche
coherent precipitate kohärentes Ausscheidungsteilchen
coherent scattering elastische Streuung
coherent twin boundary kohärente Zwillingsgrenze
coincidence site lattice (CSL) Koinzidenzgitter
cold deformation Kaltverformung
cold worked kaltverformt
colony Kolonie
color center Farbzentrum
color etching Farbätzung; Farbniederschlagätzung
columnar crystal or grain Säulenkristall; Stengelkristall
columnar structure Säulengefüge
columnar zone Säulenzone; Transkristallisationszone; Stengelkristallzone
compacted graphite verdichteter, vermikularer Graphit
compatibility diagram Kompatibilitäts-Diagram
compensating eyepiece Kompensationsokular
complex carbide Komplexkarbid; Mischkarbid
component Komponente; Bestandteil
composite Verbundwerkstoff
composition Zusammensetzung
compromise texture Kompromißtextur
Compton scattering Compton-Streuung
concentration Konzentration
condensed atmosphere kondensierte Wolke
conduction band Leitungsband
congruent kongruent
conjugate slip system konjugiertes Gleitsystem
conode Konode; Hebellinie
constitution diagram Zustandsdiagramm; Gleichgewichtsschaubild; Phasen-
    diagram
constitutional supercooling or undercooling konstitutionelle Unterkühlung
constraint Zwang

continuous cooling transformation (CCT) diagram ZTU-Schaubild/-Diagram für kontinuierliche Abkühlung

continuous grain growth stetige, kontinuierliche Kornvergrößerung; stetiges, kontinuierliches, normales Kornwachstum

continuous precipitation kontinuierliche Ausscheidung; Entmischung; kontinuierliches Zerfall

continuous recrystallization kontinuierliche Rekristallisation; *in situ*-Rekristallisation

continuous [x-ray] spectrum kontinuierliches [Röntgen]spektrum; Bremsspektrum

controlled rolling kontrolliertes Walzen

convergent beam electron diffraction (CBED) konvergente Elektronenbeugung

cooperative growth gekoppeltes Wachstum

coordination number Koordinationszahl

coordination polyhedron Koordinationspolyeder

coordination shell Koordinationssphäre

core segregation Kornseigerung; Mikroseigerung; Kristallseigerung; Dendritseigerung

coring Mikroseigerung; Kristallseigerung; Kornseigerung; Dendritseigerung

corundum Korund

Cottrell atmosphere Cottrell-Wolke

Cottrell cloud Cottrell-Wolke

coupled growth gekoppeltes Wachstum

covalent bond kovalente, homopolare Bindung

covalent crystal kovalentes Kristall

covalent radius kovalenter Radius

creep Kriechen

creep cavitation Kriechkavitation

cristobalite Cristobalit

critical cooling rate kritische Abkühlungsgeschwindigkeit

critical deformation kritischer Verformungsgrad, Reckgrad

critical point Haltepunkt; Umwandlungspunkt

critical-resolved shear stress kritische resultierende Schubspannung

critical [size] nucleus kritischer Keim

cross-slip Quergleitung

crowdion Crowdion

crystal Kristall

crystal axis Kristallachse

crystal defect Gitterbaufehler; Kristallbaufehler

crystal imperfection Gitterbaufehler; Kristallbaufehler

crystal lattice Kristallgitter; Raumgitter; Punktgitter

crystalline kristallin

crystalline anisotropy kristalline Anisotropie

crystalline ceramic kristalline Keramik

crystalline fracture kristalliner Bruch

crystallite Kristallit; Kristallkorn; Korn
crystallization Kristallisation; Erstarrung
crystallization point or temperature Kristallisationspunkt; Kristallisationstemperatur; Erstarrungspunkt; Erstarrungstemperatur
crystallographic texture [kristallographische] Textur; Vorzugsorientierung
crystal monochromator Kristallmonochromator
crystal structure Kristallstruktur
crystal system Kristallsystem
CsCl structure [type] CsCl-Gitter; Cäsiumchlorid-Strukturtyp/-Typ
CSL-boundary Koinzidenzgrenze; spezielle Korngrenze
cube-on-edge texture Goss-Lage/-Textur
cube orientation Würfellage
cube texture Würfeltextur
cubic martensite kubischer Martensit
cubic system kubisches System
Curie point or temperature (T_C, Θ_C) Curie-Temperatur (T_C, Θ_C)
curvature-driven grain growth Kornwachstum induziert durch Korngrenzenkrümmung*
Cu-type orientation Kupfer-Lage

D

δ-Fe δ-Eisen
δ-ferrite δ-Ferrit
Δr-value ΔR-Wert
dark-field illumination Dunkelfeld-Beleuchtung
dark-field image Dunkelfeldbild; Dunkelfeldabbildung
Debye–Scherrer method Debye-Scherrer-Verfahren
decomposition Entmischung; Mischkristallzerfall
decorated dislocation dekorierte Versetzung
deep center tiefes Zentrum; tiefe Störungsstelle
defect structure or lattice Fehlstruktur
deformation band Deformationsband; Transitionsband; Mikroband
deformation kinking Verformungsknicken
deformation mechanism Verformungsmechanismus
deformation mechanism map Verformungsmechanismus-Schaubild
deformation texture Verformungstextur
deformation twin Verformungszwilling
deformation twinning mechanische Zwillingsbildung; Verzwilligung
degree of freedom Freiheitsgrad
degree of long-range order Fernordnungsgrad
dendrite Dendrit; Tannenbaumkristall
dendritic segregation Dendritseigerung; Mikroseigerung; Kristallseigerung; Kornseigerung
densely packed plane Gitterebene hoher Belegungsdichte
depth of focus Tiefenschärfe

desorption Desorption
devitrification Entglasen
diamagnetic Diamagnetikum
diamond structure Diamantgitter
differential interference contrast Differenzial-Interferenzkontrast
differential scanning calorimetry (DSC) Differenzial-Rasterkalorimetrie
differential thermal analysis (DTA) Differenzialthermoanalyse
diffraction angle Beugungswinkel; Diffraktionswinkel
diffraction contrast Beugungskontrast; Orientierungskontrast
diffraction spot Beugungspunkt
diffractogram Beugungsdiagram; Interferenzdiagram; Röntgendiffraktogram
diffractometer Diffraktometer; Vielkristalldiffraktometer; Pulverdiffraktometer
diffractometric method Diffraktometerverfahren
diffuse scattering diffuse Streuung
diffusion Diffusion
diffusional creep Diffusionskriechen
diffusional plasticity Diffusionsplastizität
diffusional transformation diffusionsabhängige Umwandlung
diffusion coefficient Diffusionskoeffizient; Diffusionskonstante
diffusion-controlled diffusionsbestimmte
diffusion-induced grain boundary migration (DIGM) diffusionsinduzierte
    Korngrenzenwanderung
diffusion-induced recrystallization (DIR) diffusionsinduzierte Rekristallisa-
    tion
diffusionless transformation gekoppelte Umwandlung
diffusion porosity Kirkendall-Porosität
diffusivity Diffusionskoeffizient; Diffusionskonstante
dilation Ausdehnung
dilatometer Dilatometer
dilute [solid] solution verdünnte [feste] Lösung
direct replica einstufiger Abdruck; Filmabdruck
directional solidification gerichtete Erstarrung
disclination Disklination
discontinuous coarsening diskontinuierliche Vergröberung
discontinuous dissolution diskontinuierliche Auflösung
discontinuous grain growth anormale, unstetige, diskontinuierliche
    Kornvergrößerung; diskontinuierliches, unstetiges, anormales Kornwach-
    stum; Sekundärrekristallisation; Grobkornrekristallisation
discontinuous precipitation diskontinuierliche Ausscheidung; Zellenaus-
    scheidung
discontinuous recrystallization diskontinuierliche Rekristallisation
discontinuous yielding diskontinuierliche plastische Verformung
dislocation Versetzung
dislocation annihilation Versetzungsauflöschung; Versetzungsausheilung
dislocation core Versetzungskern
dislocation creep Versetzungskriechen

dislocation delocalization Versetzungsdelokalisation
dislocation density Versetzungsdichte
dislocation dipole Versetzungsdipole
dislocation dissociation Versetzungsaufspaltung
dislocation energy Versetzungsenergie; Versetzungslinienspannung
dislocation line tension Versetzungsenergie; Versetzungslinienspannung
dislocation loop Versetzungsschleife
dislocation multiplication Versetzungsvermehrung; Versetzungsvervielfachung
dislocation network Versetzungsnetz
dislocation pinning Versetzungsverankerung
dislocation sense Versetzungsvorzeichen
dislocation source Versetzungsquelle
dislocation splitting Versetzungsaufspaltung
dislocation stress field Versetzungsspannungsfeld
dislocation structure Versetzungsaufbau; Feingefüge
dislocation tangle Versetzungsdickicht
dislocation wall Versetzungswand
dislocation width Versetzungsbreite
disordered solid solution ungeordneter Mischkristall
disordering Fehlordnung
disorientation Misorientation
dispersed phase Dispersionsphase
dispersion strengthening Dispersionshärtung; Teilchenhärtung;
    Partikelverstärkung; Dispersionsverfestigung; Teilchenverfestigung
dispersoid Dispersoid
dispersoid-free zone dispersoidfreie Zone
displacement cascade Umlagerungsbereich; Verlagerungskaskade
displacement shift complete (DSC) lattice DSC-Gitter
displacive transformation Umklapptransformation; Schiebungsumwandlung
divacancy Doppelleerstelle
divorced eutectoid entarteter Eutektoid
divorced pearlite entarteter Perlit
dodecahedral plane Dodekaederebene
domain structure magnetische Bereichsgefüge, Bereichsstruktur;
    Domänengefüge; Domänenanordnung
domain wall Domänengrenze; Domänenwand
donor Donator
dopant Dotierstoff
doping Dotierung; Mikrolegieren
double aging Doppelaltern
double cross-slip Doppelquergleitung
double kink Doppelversetzungskinke
double stacking fault Doppelstapelfehler
doublet Dublette
drag force Hemmungskraft; rücktreibende Kraft
drift Drift-Bewegung

driving force Triebkraft; treibende Kraft
dual-phase microstructure zweiphasiges Gefüge
ductile-brittle transition [temperature] duktil-spröde Übergangstemperatur;
  Duktilitätsübergangstemperatur
ductile [cast] iron Gußeisen mit Kugelgraphit, Knotengraphit; Sphäroguß
ductility transition [temperature] Duktilitätsübergangstemperatur; duktil-
  spröde Übergangstemperatur
duplex grain size Duplexkorngröße
duplex microstructure Duplexgefüge
dynamic recovery dynamische Erholung
dynamic recrystallization dynamische Rekristallisation
dynamic strain aging dynamische Reckalterung; Portevin-Le Chatelier-Effekt

E

ε-carbide ε-Karbid
ε-martensite ε-Martensit
η-carbide η-Karbid
earing Zipfelbildung
easy glide Easy Glide
easy magnetization direction Richtung, Achse der leichtesten Magnetisierung
edge dislocation Stufenversetzung
elastic deformation elastische Verformung
elastic modulus Elastizitätsmodul; E-Modul
elastic scattering elastische Streuung
elastic strain energy Elastizitätsenergie
electro-etching elektrolytische Ätzung
electromigration Elektromigration; Elektrotransport
electron:atom ratio Valenzelektronenkonzentration
electron backscattered pattern (EBSP) Rückstreuelektronen-Beugungsbild
electron channeling Elektronenkanalierung
electron channeling pattern (ECP) Elektronen-Kanalierungsbild
electron compound or phase elektronische Verbindung, Phase; Hume–
  Rothery-Phase
electron concentration Elektronenkonzentration
electron diffraction Elektronenbeugung
electron diffraction pattern Elektronenbeugungsbild
electronegativity Elektronegativität
electron energy loss spectroscopy (EELS) Elektronenenergieverlustspek-
  troskopie
electron micrograph elektronenmikroskopische Aufnahme, Abbildung
electron [micro]probe Elektronenstrahl-Mikrosonde
electron microscopy (EM) Elektronenmikroskopie
electron probe microanalysis (EPMA) Elektronenmikrosonden-Analyse;
  Röntgen-Mikrobereichanalyse

electron spectroscopy for chemical analysis (ESCA) Elektronenspektroskopie zur chemischen Analyse; Photoelektronen-Spektroskopie
elementary jog elementarer Versetzungssprung
embryo Vorkeim
energy-dispersive diffractometry (EDS/EDAX) energiedispersive Röntgenanalyse
energy-dispersive spectrometry energiedispersive Röntgenspektroskopie
energy spectrum Energiespektrum
engineering strain Nenndehnung
engineering stress Nennspannung
epitaxial dislocation epitaktische Versetzung
epitaxial film epitaktische Dünnschicht
epitaxy Epitaxie
equatorial net Polnetz
equiaxed gleichachsig
equilibrium diagram Gleichgewichtsschaubild; Phasendiagram; Zustandsdiagram
equilibrium phase Gleichgewichtsphase; Equilibriumphase
equilibrium segregation Gleichgewichtssegregation
equilibrium system Gleichgewichtssystem
equilibrium temperature Gleichgewichtstemperatur
equivalence diagram Äquivalenz-Diagram
etch figure Ätzfigur; Kristallfigur
etch pit Ätzgrübchen
Euler angles Eulersche Winkel
eutectic colony eutektische Kolonie; eutektisches Korn
eutectic point eutektischer Punkt
eutectic reaction eutektische Reaktion
eutectic [structure] Eutektikum; eutektisches Gemisch
eutectic temperature eutektische Temperatur
eutectoid colony eutektoide Kolonie
eutectoid decomposition eutektoider Zerfall; eutektoide Entmischung
eutectoid point eutektoider Punkt
eutectoid reaction eutektoide Reaktion
eutectoid [structure] Eutektoid
eutectoid temperature eutektoide Temperatur
Ewald sphere Ewaldsche Kugel; Ausbreitungskugel
exaggerated grain growth anormales, unstetiges, diskontinuierliches, Kornwachstum; Sekundärrekristallisation; anormale, unstetige, diskontinuierliche Kornvergrößerung
excess free volume freies Exzessvolumen
exchange interaction Austauschwechselwirkung
extended dislocation aufgespaltete Versetzung
extinction Extinktion; Auslöschung
extinction coefficient Absorptionskoeffizient
extinction contour Extinktionskontur; Biegekontur; Interferenzschliere

extinction rule Auslöschungsregel
extraction replica Ausziehabdruck; Extraktionsabdruck
extrinsic grain-boundary dislocation extrinsische Korngrenzenversetzung
extrinsic stacking fault Doppelstapelfehler; extrinsischer Stapelfehler

F

face-centered cubic (FCC) structure kubisch-flächenzentrierte, kfz Struktur
face-centered lattice allseitig flächenzentriertes Gitter
F-center F-Zentrum
Fe–C system Fe–C System
Fe–Fe$_3$C system Fe–Fe$_3$C System
ferrimagnetic Ferrimagnetikum
ferrite Ferrit
ferrite-stabilizer ferritstabilisierendes Element
ferritic [cast] iron ferritisches Gußeisen
ferritic steel ferritischer Stahl
ferroelectric Ferroelektrikum
ferroelectric domain ferroelektrische Domäne
ferromagnetic Ferromagnetikum
fiber texture Fasertextur
Fick's first law 1. Ficksches Gesetz
Fick's second law 2. Ficksches Gesetz
field diaphragm Leuchtfeldblende
field emission Feldemission
field-ion microscope (FIM) Feldionenmikroskop (FIM)
field-of-view Bildfeld; Sehfeld
fine-grained feinkörnig; feinkristallin
fine pearlite Sorbit
fine structure Feingefüge; Versetzungsaufbau
firing Brennen
first-order transition Phasenumwandlung 1. Ordnung
first-order twin Primärzwilling
flake graphite Lamellengraphit
flow stress Fließspannung
fluctuation Fluktuation; Schwankung
fluorite Fluorit
fluorite [structure] type Fluorit-Strukturtyp/-Typ
forbidden gap verbotene Band; Energielücke
foreign atom Fremdatom
forest dislocation Waldversetzung
form Form
fragmentation Fragmentierung
Frank partial dislocation Franksche Teilversetzung
Frank–Read source Frank–Read-Quelle

Frank–van der Merve growth mode Frank–van-der-Merve-Wachstum[modus]
Frank vector Frank-Vektor
free energy freie Energie, Enthalpie; Helmholtzsche freie Energie
free enthalpy freie Enthalpie
Frenkel pair Frenkel-Paar
full annealing Weichglühen; Vollständigglühen
full width at half maximum (FWHM) Halbwertsbreite
fundamental translation vector Gittergrundvektor

G

γ-Fe γ-Eisen
γ'-phase γ'-Phase
garnet ferrite Ferrit-Granat
gas constant Gaskonstante
general grain boundary allgemeine Korngrenze
geometric coalescence geometrische Koaleszenz
geometrically necessary dislocations geometrisch-notwendige Versetzung*
Gibbs' free energy freie Enthalpie; Gibbssche freie Energie
Gibbs' phase rule Gibbssche Phasenregel
Gibbs–Thomson equation Gibbs-Thomson-Gleichung
glancing angle Glanzwinkel
glass-ceramic Glaskeramik
glass transition temperature (T_g) Glasübergangstemperatur (T_g)
glassy phase glasartige Substanz; amorpher Festkörper; Glasphase
glide Gleitvorgang
glissile gleitfähig
Goss texture Goss-Textur/-Lage
grain Korn; Kristallit; Kristallkorn
grain aspect ratio Kornstreckung
grain boundary Korngrenze
grain boundary allotriomorph allotriomorpher Kristall
grain-boundary character distribution Korngrenzencharakters-Verteilung*
grain-boundary diffusion Korngrenzendiffusion
grain boundary dislocation Korngrenzenversetzung
grain-boundary energy Korngrenzenenergie; Korngrenzenflächenspannung
grain-boundary mobility Korngrenzenbeweglichkeit; Korngrenzenmobilität
grain-boundary orientation Korngrenzenlage
grain-boundary segregation Korngrenzensegregation
grain-boundary sliding Korngrenzengleitung
grain-boundary strengthening Korngrenzenhärtung; Feinkornhärtung
grain-boundary tension Korngrenzenflächenspannung
grain-boundary torque Korngrenzendrehmoment
grain coalescence Kornkoaleszenz
grain coarsening Kornvergröberung

grain growth Kornwachstum; Kornvergrößerung
grain growth rate Kornwachstumsgeschwindigkeit
grain-oriented kornorientiert; texturiert
grain refining Kornfeinung
grain size Korngröße
grain size homogeneity Korngrößenhomogenität
grain size number Korngrößenkennzahl; ASTM-Korngröße
granular pearlite körniger Perlit
graphite Graphit
graphitization [annealing] Graphitisierungsglühen
graphitizer graphitstabilisierendes Element, Zusatz
gray [cast] iron graues Gußeisen; Gußeisen mit Lamellengraphit; Grauguß
green roh
Greninger–Troiano orientation relationship Orientierungsbeziehung nach
    Greninger-Troiano
groove drag Gräbchenhemmung*; Furchenhemmung*
Guinier–Preston (GP) zone Guinier-Preston-Zone; Entmischungszone

H

habit Habitus
habit plane Habitusebene
Hall–Petch equation Hall-Petch-Beziehung
hardenability Härtbarkeit
hardening [treatment] Härten
hardness Härte
Harper–Dorn creep Harper-Dorn-Kriechen
heat treatable vergütbar
heat treatment Wärmebehandlung
helical dislocation Versetzungswendel
Helmholtz free energy freie Energie
heteroepitaxial film heteroepitaktische Dünnschicht
heterogeneous microstructure heterogenes Gefüge
heterogeneous nucleation heterogene Keimbildung
heterogeneous system heterogenes System
heterojunction Heteroübergang
heterophase mehrphasig
heteropolar bond heteropolare Bindung; Ionenbindung
heterostructure Heterostruktur
hexagonal close-packed (HCP) structure hexagonal-dichtestgepackte/hdp
    Struktur
hexagonal ferrite Hexaferrit
hexagonal system hexagonales System
high-angle grain boundary Großwinkelkorngrenze
high-resolution transmission electron microscope (HRTEM) Hochauflö-
    sungs-Elektronenmikroskop

high-temperature thermo-mechanical treatment hochtemperatur–thermomechanische Behandlung
high-voltage electron microscope (HVEM) Hochspannungs-Elektronenmikroskop
homoepitaxial film homoepitaktische Dünnschicht
homogeneous microstructure homogenes Gefüge
homogeneous nucleation homogene Keimbildung
homogeneous system homogenes System
homogenizing [anneal] or homogenization Diffusionsglühen; Homogenisierung
homologous temperature homologische Temperatur
homopolar bond homopolare, kovalente Bindung
Hooke's law Hookesches Gesetz
host atom Wirtsatom; Matrixatom
hot deformation Warmverformung
hot isostatic pressing (HIP) heißisostatisches Pressen
hot pressing Heißpressen
hot worked warmverformt
hot-stage microscope Hochtemperaturmikroskop; Heiztischmikroskop
Hume–Rothery phase Hume–Rothery-Phase; elektronische Verbindung, Phase
Hume–Rothery rules Hume–Rothery-Regeln
hydrostatic pressure hydrostatischer, allseitiger Druck
hypereutectic übereutektisch
hypereutectoid übereutektoid
hypoeutectic untereutektisch
hypoeutectoid untereutektoid

I

ideal orientation Ideallage
immersion objective or lens Immersionsobjektiv
imperfect dislocation Teilversetzung; unvollständige Versetzung; Partialversetzung
impurity Verunreinigung; Beimengung
impurity cloud Fremdatomenwolke
impurity drag Verunreinigungshemmung*
incoherent interface inkohärente Grenzfläche
incoherent precipitate or particle inkohärentes Ausscheidungsteilchen
incoherent twin boundary inkohärente Zwillingsgrenze
incubation period Inkubationszeit
indirect replica zweistufiger Abdruck; Matrizenabdruck
induction period Inkubationszeit
inelastic scattering unelastische Streuung
ingot Block
inhomogeneous microstructure inhomogenes Gefüge
inoculant Impfstoff

in situ **observation** *in situ*-Beobachtung
instrumental [x-ray] line broadening instrumentale Linienverbreiterung
integral [x-ray] line width Integrallinienbreite
integrated [x-ray] line intensity Integral[linien]intensität
interatomic spacing Atomabstand; Gitterpunktabstand
intercritical heat treatment interkritische Wärmebehandlung
intercritical range interkritischer Umwandlungsbereich
intercrystalline interkristallin
interdiffusion Interdiffusion; chemische Diffusion
interface Grenzfläche; Phasengrenze
interface-controlled phasengrenzenbestimmte
interfacial energy Phasengrenzenenergie
intergranular interkristallin
interlamellar spacing Lamellenabstand
intermediate phase Zwischenphase; intermediäre Phase
intermetallic compound Metallid; intermetallische Verbindung
internal friction innere Reibung; Dämpfung
internal oxidation innere Oxidation
internal stresses innere Spannungen; Eigenspannungen
interphase precipitation Interphasenausscheidung
interplanar spacing Netzebenenabstand
interstice Zwischengitterplatz; Gitterlücke
interstitial fremdes Zwischengitteratom; interstitielles Fremdatom
interstitial compound Einlagerungsphase; interstitielle Phase
interstitialcy eigenes Zwischengitteratom
interstitial [mechanism of] diffusion Zwischengitterdiffusion; interstitielle
  Diffusion
interstitial [foreign] atom fremdes Zwischengitteratom; interstitielles Fremdatom
interstitial phase interstitielle Phase; Einlagerungsphase
interstitial solid solution Einlagerungsmischkristall; interstitieller Mischkristall
intracrystalline intrakristallin; transkristallin
intragranular intrakristallin; transkristallin
intrinsic diffusion coefficient intrinsische Diffusionskonstante
intrinsic diffusivity intrinsische Diffusionskonstante
intrinsic grain-boundary dislocation intrinsische Korngrenzenversetzung
intrinsic stacking fault intrinsischer Stapelfehler
intrinsic [x-ray] line broadening physikalische Linienverbreiterung
invariant reaction invariante Reaktion
inverse pole figure reziproke Polfigur
ion channeling Ionenkanalierung
ion etching Ionenätzung
ionic bond Ionenbindung; heteropolare Bindung
ionic crystal Ionenkristall
ionic radius Ionenradius

ion implantation Ionenimplantieren
irradiation damage Strahlenschäden
irradiation defects strahleninduzierte Kristallbaufehler; Bestrahlungsdefekte
irradiation growth strahleninduzierter Wachstum
irradiation hardening strahleninduzierte Aushärtung
irradiation-induced creep strahleninduziertes Kriechen
irreversible unumkehrbar; irreversibel
island film Inseldünnschicht
isochronal annealing isochrones Glühen
isoforming Isoforming
isomorphism Isomorphie
isomorphous phases isomorphe Phasen
isomorphous system isomorphes System
isothermal isotherm
isothermal transformation isotherme Umwandlung
isothermal transformation diagram isothermes Zeit-Temperatur-Umwand-
    lungs-/ZTU-Schaubild; ZTU-Diagram
isotropic isotrop

J

jog Versetzungssprung
Johnson–Mehl–Kolmogorov equation Gleichung von Johnson-Mehl-Kolmog-
    orov

K

Kelvin's tetrakaidecahedron Kelvin-Tetrakaidekaeder
Kê peak or relaxation Kê-Dämpfungsmaximum/-Relaxation
Kerr microscopy Kerr-Mikroskopie
Kikuchi lines Kikuchi-Linien
kinetics [of transformation] Umwandlungskinetik
kink Versetzungskinke
kink band Knickband
Kirkendall effect Kirkendall-Effekt
Kossel line pattern Kossel-Linienbild
Köster effect Köster-Effekt
Köster peak or relaxation Snoek-Köster-Dampfungsmaximum/-Relaxation
Kurdjumov–Sachs orientation relationship Orientierungsbeziehung nach
    Kurdjumov-Sachs

L

laminar slip laminare Gleitung
Lankford coefficient Lankford-Wert; $\bar{r}$-Wert

large-angle grain boundary Großwinkelkorngrenze
Larson–Miller parameter Larson-Miller-Parameter
latent hardening latente Verfestigung
lath martensite lanzettförmiger, massiver Martensit; Blockmartensit
lattice Raumgitter; Gitter
lattice basis Gitterbasis
lattice constant Gitterkonstante; Gitterperiode
lattice diffusion Volumendiffusion
lattice direction Gittergerade
lattice-matched [epitaxial] film gitter-anpassende [epitaktische] Dünnschicht*
lattice misfit Gitterfehlpassung
lattice-mismatched [epitaxial] film gitter-fehlpassende [epitaktische]
    Dünnschicht*
lattice parameters Gitterparameter
lattice plane Gitterebene; Netzebene
lattice point Gitterpunkt; Gitterknoten
lattice site Gitterknoten; Gitterpunkt
lattice void Zwischengitterplatz; Gitterhohlraum; Gitterlücke
Laue diffraction pattern Laue-Aufnahme
Laue equations Laue-Gleichungen/-Bedingungen
Laue method Laueverfahren; Lauemethode
Laves phase Laves-Phase
ledeburite Ledeburit
ledeburitic steel ledeburitischer Stahl
lenticular martensite linsenförmiger Martensit; Plattenmartensit
lever rule Hebelbeziehung; Konodenregel
lineage structure Zellensubgefüge
linear absorption coefficient linearer Schwächungskoeffizient
linear defect linearer Defekt; linienförmiger Gitterfehler
linear growth rate lineare Wachstumsgeschwindigkeit
line broadening Linienverbreiterung
liquation Seigerung; Makroseigerung; Blockseigerung; Stückseigerung
liquid crystal Flüssigkristall
liquid-phase sintering Flüssigphasensintern
liquidus Liquidus[kurve]; Liquidus[linie]
logarithmic creep logarithmisches Kriechen
Lomer–Cottrell barrier or lock Lomer-Cottrell-Schwelle
long-range weitreichend
long-range order Fernordnung
long-range ordering Ordnungsumwandlung
long-range order parameter Fernordnungsparameter
Lorentz factor Lorentz-Faktor
Lorentz microscopy Lorentz-Mikroskopie
low-angle boundary Kleinwinkelkorngrenze; Subkorngrenze
low-energy electron diffraction (LEED) Niedrigenergetischenelektronenbeu-
    gung

lower bainite unterer, nadeliger Bainit
lower yield stress untere Streckgrenze
low-temperature thermo-mechanical treatment Austenitformlungshärten;
tieftemperatur-thermomechanische Behandlung
Lüders band Lüdersscher Band
Lüders strain Lüderssche Dehnung

M

M_d **temperature** M_d-Temperatur
M_s^σ **temperature** M_s^σ-Temperatur
macrograph Makroaufnahme
macroscopic stress Eigenspannung 1. Art; Makroeigenspannung
macrosegregation Makroseigerung; Blockseigerung; Stückseigerung;
Seigerung
macrostructure Grobgefüge; Makrogefüge
magnetic crystalline anisotropy magnetische Kristallanisotropie
magnetic domain magnetischer Bezirk; Domäne; Weißscher Elementarbereich
magnetic force microscope (MFM) Magnetkraft-Mikroskop
magnetic ordering magnetische Ordnung
magnetic structure magnetische Bereichstruktur; Domänenstruktur
magnetic texture magnetische Textur
magnetic transformation magnetische Unwandlung
magnification Vergrößerung; Abbildungsmaßstab
major segregation Makroseigerung; Blockseigerung; Stückseigerung;
Seigerung
malleable [cast] iron Temperguß
maraging steel martensitaushärtender Stahl
marquenching isothermes Härten
martempering isothermes Härten
martensite Martensit
martensite finish temperature (M_f) Temperatur vom Ende der Martensitbil-
dung* (M_f-Temperatur)
martensite start temperature (M_s) Temperatur vom Anfang der Martensitbil-
dung* (M_s-Temperatur)
martensitic range Martensit-Temperaturbereich/-Temperaturintervall
martensitic steel martensitscher Stahl
martensitic transformation Martensitumwandlung
mass absorption coefficient Massenschwächungskoeffizient
massive martensite massiver, lanzettförmiger Martensit; Blockmartensit
massive transformation massive Umwandlung
master alloy Vorlegierung
matrix Matrix; Grundmasse
matrix band Matrixband
Matthiessen rule Matthiessensche Regel
M-center M-Zentrum

mean [grain/particle] size mittlere, durchschnittliche Größe
mechanical alloying mechanisches Legieren
mechanical anisotropy mechanische Anisotropie
mechanical property mechanische Eigenschaft; Festigkeitseigenschaft
mechanical stabilization [of austenite] mechanische Stabilisation [von Austenit]
median size Mediangröße
melt spinning Schmelzspinnverfahren
melting point/temperature (T_m) Schmelzpunkt; Schmelztemperatur
mesomorphic phase or mesophase Mesophase
metadynamic recrystallization metadynamische Rekristallisation
metal ceramic Metallkeramik
metallic bond Metallbindung
metallic crystal metallisches Kristall
metallic glass metallisches Glas
metallic radius metallischer Radius
metallographic examination metallographische Untersuchung
metallographic section or sample [metallographischer] Schliff; Anschliff
metastable β alloy metastabile β–Legierung
metastable β-phase ($β_m$) metastabile β-Phase ($β_m$)
metastable phase metastabile Phase
metastable state metastabiler Zustand
microalloying Mikrolegieren; Dotierung
microanalysis Mikroanalyse
microband Deformationsband; Mikroband
microconstituent Gefügebestandteil; Gefügeelement
microdiffraction Feinbereichbeugung
micrograph Mikroaufnahme
microprobe Elektronenstrahlmikrosonde
microscopic stress Eigenspannung 2. Art; Mikroeigenspannung
microsegregation Kornseigerung; Kristallseigerung
microstrain Mikrodehnung; Mikroverformung
microstructure Mikrogefüge; Mikrostruktur
microtexture Mikrotextur
midrib Mittelrippe
Miller indices Millersche Indizes
Miller–Bravais indices Miller-Bravais-Indizes
mirror plane Spiegelebene
miscibility gap Mischungslücke
misfit dislocation Fehlpassungsversetzung; Phasengrenzenversetzung
misfit parameter Fehlpassungsparameter
misorientation Misorientation
misorientation distribution function Misorientations-Verteilungsfunktion
mixed dislocation gemischte Versetzung
mixed grain boundary gemischte Korngrenze
mode Modalgröße; Modus

modification Modifizierung; Veredelung
modulated structure moduliertes Gefüge; Tweedgefüge
modulus of elasticity Elastizitätsmodul; E-Modul
modulus of rigidity Schermodul; Schubmodul
moiré pattern Moiré-Bild
mol% Molarkoncentration; Mole-%
monochromatic radiation monochromatische Strahlung
monoclinic system monoklines System
monocrystalline einkristallin
monotectic reaction monotektische Reaktion
monotectoid reaction monotektoide Reaktion
mosaic structure Mosaikgefüge
most probable size Modalgröße; Modus
multiple cross-slip mehrfache Quergleitung
multiple jog mehrfacher Versetzungssprung
multiple slip mehrfache Gleitung
multiplicity factor Flächenhäufigkeitsfaktor

N

Nabarro–Herring creep Nabarro-Herring-Kriechen
NaCl structure Natriumchloridgitter; Steinsalzgitter
nanocrystalline nanokristallin
natural aging Kaltauslagerung; Kaltaushärtung
N crystal nematischer Kristall
N* crystal cholesterischer Kristall
nearly special grain boundary fast-spezielle Korngrenze
Néel point or temperature (T_N, Θ_N) Néel-Temperatur (T_N, Θ_N)
Néel wall Néel-Wand
nematic crystal nematischer Kristall
net plane Netzebene; Gitterebene
Neumann band Neumannscher Band
neutron diffraction Neutronenbeugung
n-fold axis n-zählige Drehachse
Nishiyama orientation relationship Orientierungsbeziehung nach Nishiyama-
      Wassermann
nitride Nitrid
nodular [cast] iron Gußeisen mit Knotengraphit, Kugelgraphit
nodular graphite Knötchengraphit; Kugelgraphit; sphärolytscher Graphit
nominal strain technische Dehnung
nominal stress technische Spannung; Nennspannung
nondiffusional transformation gekoppelte Umwandlung
non-oriented regellos-orientiert
normal anisotropy senkrechte Anisotropie
normal grain growth normales, kontinuierliches, stetiges Kornwachstum; ste-
      tige, kontinuierliche Kornvergrößerung

normalizing Normalglühen; Normalisieren
normal stress normale Spannung
nucleation Keimbildung
nucleation agent Keimbildner
nucleation rate Keimbildungsgeschwindigkeit
nucleus Keim
numerical aperture numerische Apertur

O

octahedral interstice Oktaeder-Zwischengitterplatz; Oktaedergitterlücke; Oktaederhohlraum
octahedral plane Oktaederebene
octahedral site Oktaeder-Zwischengitterplatz; Oktaedergitterlücke; Oktaederhohlraum
octahedral void Oktaederhohlraum; Oktaeder-Zwischengitterplatz; Oktaedergitterlücke
one-way shape memory effect Einweg-Formgedächtniseffekt
optical microscope Lichtmikroskop; Auflichtmikroskop; Metallmikroskop
orange peel Orangenhaut
order–disorder transformation or transition Ordnungsumwandlung; Ordnungs-Unordnungs-Umwandlung
ordered solid solution ferngeordneter Mischkristall
orientation distribution function (ODF) Orientierungsverteilungsfunktion (OVF)
orientation imaging microscopy (OIM) orientierungsabbildende Mikroskopie
orientation relationship Orientierungsbeziehung; Orientierungszusammenhang
orientation sphere Lagekugel
orientation spread Texturschärfe
Orowan loop Orowan-Schleife
Orowan mechanism Orowan-Mechanismus
orthoferrite Orthoferrit
orthorhombic system [ortho]rhombisches System
Ostwald ripening Ostawldsche Reifung; Umlösung
overaging Überalterung
oxide dispersion strengthened (ODS) oxidendispersionsgehärtet
oxynitride Oxynitrid

P

packet martensite lanzettförmiger, massiver Martensit; Blockmartensit
packing factor Packungsdichte; Packungsfaktor; Raumerfüllung
paramagnetic Paramagnetikum
partial dislocation unvollständige Versetzung; Partialversetzung; Teilversetzung

partially coherent interface teilkohärente Phasengrenze
partially coherent precipitate teilkohärentes Präzipitat, Teilchen
partially ordered solid solution teilgeordneter Mischkristall
particle coarsening Teilchenvergröberung
particle drag Teilchenhemmung; Ausscheidungshemmung
particle shearing Teilchen-Schneidemechanismus*
particle-stimulated nucleation (PSN) Rekristallisationskeimbildung an
     Teilchen*
pearlite Perlit
pearlitic cast iron perlitischer Gußeisen
pearlitic colony or nodule Perlitkolonie; Perlitkorn
pearlitic range Perlit-Temperaturbereich/-Temperaturintervall
pearlitic steel perlitischer Stahl
pearlitic transformation Perlit-Umwandlung
Peierls stress or barrier Peierls-Spannung/-Schwelle
pencil glide Pencile Glide; Prismengleitung
perfect dislocation vollständige Versetzung
peritectic reaction peritektische Reaktion
peritectic temperature peritektische Temperatur
peritectoid reaction peritektoide Reaktion
perovskite Perowskit
perovskite [structure] type Perowskit-Strukturtyp/-Typ
phase Phase
phase boundary Phasengrenze
phase composition Phasenbestand
phase constituent Phasenbestandteil
phase contrast Phasenkontrast
phase diagram Gleichgewichtsschaubild; Zustandsdiagram; Phasendiagram
phase rule Phasenregel; Phasengesetz
phase transformation or transition Phasenumwandlung; Phasenübergang
photo-electron emission microscope (PEEM) Fotoemissions-Elektronenmik-
     roskop
physical adsorption physikalische Adsorption
physical property physikalische Eigenschaft
physisorption physikalische Adsorption
piezoelectric Piezoelektrikum
pile-up Versetzungsaufstau
pinning force hemmende, rückhaltende Kraft; Hemmungskraft
pipe diffusion Pipe Diffusion
plain carbon steel unlegierter Stahl; Kohlenstoffstahl
planar anisotropy Flächenanisotropie
planar defect flächenförmiger Gitterfehler
plastic deformation plastische Verformung
plate martensite nadelförmiger, linsenförmiger Martensit; Plattenmartensit;
     Lattenmartensit
point defect Punktdefekt; Punktfehler; atomare Fehlstelle

point group Punktgruppe; Punktsymmetriegruppe; Symmetrieklasse
point lattice Punktgitter; Kristallgitter; Raumgitter
Poisson's ratio Poissonsche Zahl; Querkontraktionszahl
polar net Polnetz
polarization factor Polarisationsfaktor
polarized-light microscopy Polarisationslichtmikroskopie
pole Pol
pole figure Polfigur
polychromatic [x-ray] radiation Bremsstrahlung; polychromatische
    Röntgenstrahlung
polycrystal Vielkristall
polycrystalline vielkristallin
polygonization Polygonisation
polygonized polygonisiert
polymorphic crystallization polymorphe Kristallisation*
polymorphic modification polymorphe Modifikation
polymorphic transformation polymorphe Umwandlung
polymorphism Polymorphie
polytypism Polytypie
porosity Porosität
Portevin–Le Chatelier effect Portevin-Le Chatelier-Effekt; dynamische Reck-
    alterung
postdynamic recrystallization postdynamische Rekristallisation
powder method Pulververfahren
powder pattern Pulverdiagram
power-law creep exponentielles Kriechen
precipitate Präzipitat; Ausscheidung
precipitated phase Ausscheidungsphase; Segregat
precipitate reversion Rückbildung
precipitation Ausscheidung; Segregatbildung
precipitation-free zone (PFZ) teilchenfreie Zone
precipitation strengthening or hardening Aushärtung; Ausscheidungshär-
    tung; Alterungshärtung
precipitation treatment Ausscheidungsbehandlung
preferred grain orientation Vorzugsorientierung; ausgeprägte Kornorien-
    tierung; Textur
preformed nucleus vorgebildeter Keim
preprecipitation Vor-Ausscheidung
primary α-phase [in Ti alloys] primäre Alpha-Phase
primary creep primäres Kriechen; Übergangskriechen
primary crystals Primärkristallen
primary dislocation primäre Versetzung
primary extinction Primärextinktion
primary recrystallization Primärrekristallisation
primary slip system Hauptgleitsystem
primary solid solution Endmischkristall; Primärmischkristall

primary structure Primärgefüge
primitive lattice primitives, einfaches Raumgitter, Gitter
prismatic [dislocation] loop prismatische Versetzungsschleife
prismatic slip Prismengleitung
prism plane Prismenebene
proeutectoid voreutektoid
proeutectoid cementite voreutektoider Zementit; Sekundärzementit
proeutectoid ferrite voreutektoider, untereutektoider Ferrit
pseudoplasticity Pseudoplastizität
pyramidal plane Pyramidenebene
pyramidal slip Pyramidengleitung

Q

quantitative metallography quantitative Metallographie; Stereologie
quartz Quarz
quasi-crystal Quasikristall
quasi-isotropic quasiisotrop
quench aging Abschreckalterung
quench hardening Abschreckhärtung
quench-in vacancy eingeschreckte Leerstelle
quenching Härten; Abschreckung

R

radial distribution function (RDF) Radialverteilungsfunktion*
radiation damage Strahlenschaden
random grain boundary allgemeine Korngrenze
random solution ungeordneter Mischkristall
R-center R-Zentrum
reciprocal lattice reziprokes Gitter
reconstructive transformation rekonstruktive Umwandlung
recovery Erholung; Ausheilung
recrystallization (ReX) Rekristallisation
recrystallization annealing Rekristallisationsglühen
recrystallization diagram Rekristallisationsdiagram
recrystallization *in situ* *in situ*-Rekristallisation; kontinuierliche Rekristallisation
recrystallization nucleus Rekristallisationskeim
recrystallization temperature Rekristallisationstemperatur
recrystallization texture Rekristallisationstextur
recrystallized rekristallisiert
reference sphere Lagekugel
reflection high-energy electron diffraction (RHEED) RHEED
reflection sphere Ewaldsche Kugel; Reflexionskugel; Ausbreitungskugel

relaxation modulus relaxierter Modul
relaxation time Relaxationszeit
replica Abdruck
residual austenite Restaustenit
residual electrical resistance Restwiderstand
residual stresses Eigenspannungen; Restspannungen
resolution limit Auflösungsgrenze
resolved shear stress resultierende Schubspannung
resolving power Auslösungsvermögen
retained austenite Restaustenit
retained β-phase Rest-Beta-Phase
retrograde solidus retrograde Soliduskurve
reversibility Reversibilität
reversible temper brittleness reversible Anlassversprödung
reversion Rückbildung
rhombohedral system rhomboedrisches, trigonales System
rocking curve Reflexionskurve; Rocking-Kurve
rock salt [structure] type Steinsalz-Strukturtyp/-Typ
Rodrigues vector Rodrigues-Vektor
roller quenching Schmelzspinn-Verfahren
R-orientation R-Lage
rotating crystal method Drehkristallverfahren
rule of stages Stufenregel
r-value r-Wert
$\bar{r}$-value $\bar{r}$-Wert; Lankford-Wert

S

σ-plot Wulff-Plot
Sachs factor Sachsscher Faktor
sample thickness effect Probendickeneffekt*
saturated solid solution gesättigter Mischkristall
S/Bs-orientation S/Ms-Lage
scanning Auger-electron microscope (SAM) Auger-Rastermikrosonde
scanning electron microscope (SEM) Rasterelektronenmikroskop (REM)
scanning transmission electron microscope (STEM) Raster-Durchstrah-
    lungs-Elektronenmikroskop
scanning tunneling microscope (STM) Raster-Tunnel-Elektronenmikroskop
    (RTM)
Schmid factor Schmidsches Faktor
Schmid's law Schmidsches Gesetz
Schottky pair Schottky-Paar
screw dislocation Schraubenversetzung
S crystal smektischer Kristall
secondary cementite Sekundärzementit; voreutektoider Zementit
secondary creep sekundäres, stationäres Kriechen

secondary crystals Sekundärkristalle
secondary dislocation sekundäre Versetzung
secondary electron Sekundärelektron
secondary extinction Sekundärextinktion
secondary hardening Sekundärhärtung
secondary ion mass spectroscopy (SIMS) Sekundärionen-Massenspektroskopie
secondary precipitate sekundäre Ausscheidung
secondary recrystallization Sekundärrekristallisation; anormale, unstetiges, diskontinuierliches Kornwachstum; anormale, unstetige, diskontinuierliche Kornvergrößerung; Grobkornrekristallisation
secondary slip system sekundäres Gleitsystem
secondary structure Sekundärgefüge
second-order transition Phasenumwandlung 2. Ordnung
second-order twin Sekundärzwilling*
second phase sekundäre Phase
seed crystal Impfkristall
segregation Seigerung
selected area channeling pattern (SACP) Feinbereichs-Kanalierungsbild
selected area diffraction (SAD/ESAD) Feinbereichsbeugung
self-diffusion Selbstdiffusion
self-interstitial eigenes Zwischengitteratom
self-similar skaleninvariant; selbstähnlich
semi-coherent interface teilkohärente Phasengrenze
sessile dislocation nicht-gleitfähige, sessile Versetzung
shadowing [Schräg]Beschattung; Bedampfung
shallow impurity flache Störungsstelle
shape memory effect Formgedächtniseffekt
sharp yield point ausgeprägte Streckgrenze
shear Schubbeanspruchung
shear band Scherband
shear modulus Schubmodul; Gleitmodul
shear strain Schiebung
shear stress Schubspannung
shear[-type] transformation Schiebungsumwandlung; Umklappumwandlung
sheet texture Blechtextur
Shockley partial dislocation Shockley-Teilversetzung
short-circuit diffusion path Pfad, Weg der bevorzugten Diffusion*
short-range kurz-reichend
short-range order Nahordnungszustand
short-range ordering Nahordnung
short-range order parameter Nahordnungsparameter
shrinkage Schwindung; Volumenkontraktion
sialon Sialon
silica Siliciumdioxid
simple lattice einfaches, primitives Gitter/Raumgitter

single crystal Einkristall
single-domain particle Eindomänen-Teilchen
single slip einfache Gleitung
sintering Sintern
size distribution Größenverteilung
slip Gleitvorgang; Gleitung
slip band Gleitband
slip direction Gleitrichtung
slip line Gleitlinie
slip plane Gleitebene
slip system Gleitsystem
slip trace Gleitspur
small-angle grain boundary Kleinwinkelkorngrenze
smectic crystal smektischer Kristall
Snoek–Köster peak/relaxation Snoek–Köster-Dämpfungsmaximum/-Relaxation
Snoek peak or relaxation Snoek-Dämpfungsmaximum/-Relaxation
soaking Durchwärmen
softening anneal Weichglühen
solid fest
solidification Erstarrung
solidification point or temperature Erstarrungspunkt; Erstarrungstemperatur; Gefrierpunkt
solid solubility Löslichkeit im festen Zustand
solid solution Mischkristall; feste Lösung
solid solution strengthening or hardening Mischkristallverfestigung; Mischkristallhärtung
solid-state sintering Festphasensintern
solidus Solidus[kurve]; Solidus[linie]
solubility limit Löslichkeitsgrenze
solute aufgelöster Stoff
solute diffusion Fremdatomendiffusion
solute drag Fremdatomenhemmung*
solution treatment Lösungsbehandlung
solvent Matrixstoff; Wirtsstoff
solvus Solvus [kurve]; Löslichkeits [linie]
sorbite Sorbit
S-orientation S-Ideallage
sorption Sorption
space group Raumgruppe
special carbide Sonderkarbid
special grain boundary spezielle Korngrenze; Koinzidenzgrenze
specific [interface] area spezifische Grenzenfläche
specific volume spezifisches Volumen
sphalerite [structure] type Sphalerit-Strukturtyp/-Typ
sphere of reflection Ewaldsche Kugel; Ausbreitungskugel

spherical aberration sphärische Abberation; Öffnungsfehler
spheroidal graphite Kugelgraphit; sphärolitscher Graphit; Knötchengraphit
spheroidite körniger, globularer, kugeliger Perlit
spheroidization Kugelbildung
spheroidized pearlite körniger, globularer Perlit
spherulite Sphärolyt
spinel Spinell
spinel ferrite Ferrit-Spinell
spinel [structure] type Spinell-Strukturtyp/-Typ
spinodal Spinodale
spinodal decomposition spinodale Entmischung
spontaneous spontan
sputtering Zerstäubung
stabilized ZrO₂ stabilisiertes Zirkoniumdioxid
stable phase Gleichgewichtsphase; Equilibriumphase
stacking fault Stapelfehler
stacking-fault energy (SFE) Stapelfehlerenergie
staining Anlassätzung
stair-rod dislocation Kantenversetzung
standard (*hkl*) projection Standard-(*hkl*)-Projektion
standard triangle Standarddreieck; Grunddreieck
static lattice distortion statische Gitterverzerrung; Eigenspannung 3. Art; Gitterstörung
static recovery statische Erholung
static recrystallization statische Rekristallisation
steady-state creep stationäres, sekundäres Kriechen
steel Stahl
steel martensite Stahl-Martensit
stereographic net stereographisches Netz
stereographic projection stereographische Projektion
stereology Stereologie; quantitative Metallographie
stoichiometry or stoichiometric [composition] Stöchiometrie
stored energy gespeicherte Verformungsenergie
strain Verformungsgrad; Formänderung
strain aging Verformungsalterung
strained-layer epitaxy Strained-Layer-Epitaxie
strain hardening Kaltverfestigung; Verformungsverfestigung
strain-hardening exponent Verfestigungsexponente
strain-induced grain boundary migration (SIBM) verformuungsinduzierte Korngrenzenwanderung*
strain-induced martensite Verformungsmartensit
strain rate Formänderungsgeschwindigkeit; Verformungsgeschwindigkeit
strain rate sensitivity Empfindlichkeit zur Formänderungsgeschwindigkeit
strain ratio r-Wert
Stranski–Krastanov growth mode Stranski–Krastanov-Wachstum[modus]
stress Spannung

stress-assisted martensite spannungsinduzierter Martensit
stress–deformation diagram Spannungs-Dehnungs-Diagram
stress-induced martensite spannungsinduzierter Martensit
stress relaxation Entspannung; Relaxation
stress-relief anneal Entspannungsglühen; Spannungsarmglühen; Spannungsab-
    bauglühen
stress–strain diagram wahre Spannungs-wahre Dehnungs-Kurve; Fließkurve
stretcher-strain marking Fließfigur; Kraftwirkungsfigur
striation structure Zellengefüge
structural disorder strukturelle Fehlordnung
structural vacancy strukturelle Leerstelle
structure Struktur; Gefüge
structure factor or amplitude Strukturfaktor; Strukturamplitude
structure-insensitive gefüge-unabhängig
structure-sensitive gefüge-abhängig
subboundary Subkorngrenze; Kleinwinkelkorngrenze
subcritical annealing subkritisches Glühen
subgrain Subkorn
subgrain boundary Subkorngrenze; Kleinwinkelkorngrenze
subgrain coalescence Subkornkoaleszenz
subgrain structure Subkorngefüge
sublattice Teilgitter; Untergitter
substitutional atom Austauschatom; Substitutionsatom
substitutional solid solution Austauschmischkristall; Substitutionsmisch-
    kristall
substructure Feingefüge; Subgefüge
supercooling Unterkühlung
superdislocation Überstrukturversetzung
superheating Überhitzung
superlattice Überstruktur
superplasticity Superplastizität
supersaturation Übersättigung
superstructure Überstruktur
surface-energy driving force Oberflächenenergie-Triebkraft*
surface tension Oberflächenspannung
Suzuki atmosphere Suzuki-Atmosphäre
symmetric boundary symmetrische Korngrenze
symmetry axis Symmetrieachse; Gire
symmetry class Symmetrieklasse; Punktgruppe; Punktsymmetriegruppe
symmetry element Symmetrieelement
symmetry operation Symmetrieoperation
system System

T

Taylor factor Taylor-Faktor/-Orientierungsfaktor

temper carbon Temperkohle
tempered martensite angelassener Martensit
tempering of steel martensite Anlassen von Stahl-Martensit
tempering of titanium martensite Anlassen von Titan-Martensit
tempering [treatment] Anlassen; Tempern
temper rolling Dressieren; Nachwalzen
tensile strain Dehnung
tensile stress Zugspannung
tension Zugbeanspruchung; Zugumformung
terminal solid solution Endmischkristall; Primärmischkristall
ternary dreistoff
tertiary cementite tertiärer Zementit
tertiary creep tertiäres Kriechen
tertiary recrystallization tertiäre Rekristallisation
tetragonal system tetragonales System
tetragonality Tetragonalität
tetrahedral interstice Tetraeder-Zwischengitterplatz; Tetraedergitterlücke; Tetraederhohlraum
tetrahedral site Tetraeder-Zwischengitterplatz; Tetraedergitterlücke; Tetraederhohlraum
tetrahedral void Tetraederhohlraum; Tetraedergitterlücke; Tetraeder-Zwischengitterplatz
tetrakaidecahedron Tetrakaidekaeder
texture [kristallographische] Textur; Vorzugsorientierung
texture analysis Texturanalyse
texture component Ideallage; Vorzugslage
texture intensity Texturbelegung
texture scatter Texturschärfe
theoretical strength theoretische Festigkeit
thermal analysis thermische Analyse
thermal etching thermische Ätzung; Vakuumätzung
thermal groove thermisches Gräbchen; thermische Furche; Oberflächenfurche
thermal hysteresis thermische Hysterese
thermally activated thermisch-aktiviert
thermally hardened thermisch-aushärtet
thermal stability thermische Stabilität
thermal stresses thermoelastische Spannungen
thermal treatment Wärmebehandlung
thermodynamic equilibrium thermodynamisches Gleichgewicht
thermodynamic stability thermodynamische Stabilität
thermoelastic martensite thermoelastischer Martensit
thermo-magnetic treatment thermomagnetische Behandlung
thermo-mechanical processing thermomechanische Behandlung
thermo-mechanical treatment thermomechanische Behandlung
thickness fringes Dickenkonturen; Keilinterferenzen; Streifenkontrast
thin foil Folie

Thompson's tetrahedron Thompson-Tetraeder
Thomson–Freundlich equation Thomson-Freundlich-Gleichung
tie line Konode; Hebellinie
tilt grain boundary Kippgrenze
time-temperature-transformation (TTT) diagram Zeit-Temperatur-
    Umwandlungs-/ZTU-Schaubild; ZTU-Diagram
tinting Farbätzung; Farbniederschlagsätzung
titanium martensite Titan-Martensit
transcrystalline transkristallin; intrakristallin
transcrystallization zone Transkristallisationszone; Säulenzone; Stengelkri-
    stallzone
transformation hysteresis Umwandlungshysterese
transformation-induced plasticity (TRIP) umwandlungsinduzierte Plastizität
transformation range Umwandlungstemperaturbereich
transformation rate Umwandlungsgeschwindigkeit
transformation stresses Umwandlungsspannungen
transformation toughening Umwandlungsverstärkung
transformation twin Umwandlungszwilling
transformed β structure (β_{tr}) umgewandelte Beta-Gefüge
transgranular transkristallin
transient creep Übergangskriechen; primäres Kriechen
transient phase Zwischenphase
transition band Transitionsband; Deformationsband; Mikroband
transition phase Zwischenphase
translation group Translationsgruppe
translation vector Translationsvektor; Translationsperiode
transmission electron microscope (TEM) Durchstrahl-, Transmissions-Elek-
    tronenmikroskop (TEM)
transmission Laue method Laue-Transmissionsverfahren
transus Gleichgewichtslinie
triclinic system triklines System
tridymite Tridymit
trigonal system trigonales, rhomboedrisches System
triple junction Tripelpunkt; Kornkante
triple point Tripelpunkt
trostite Troostit
true strain wahre Dehnung, Verformung
true stress wahre Spannung
twin Zwilling
twinned crystal Zwillingskristall
twinning system Zwillingsbildungssystem
twist disclination Schraubendisklination
twist grain boundary Drehkorngrenze
two-phase structure zweiphasiges Gefüge
two-way shape memory effect Zweiweg-Formgedächtniseffekt

U

ultra-fine grained ultrafeinkörnig; submikrokristallin
undercooling Unterkühlung
unit cell Elementarzelle; Einheitszelle
unit cell parameter Elementarzellenparameter
unit stereographic triangle Standarddreieck
unsaturated [solid] solution nichtgesättigter Mischkristall
uphill diffusion Bergauf-Diffusion
upper bainite oberer, körniger Bainit
upper yield stress obere Fließgrenze

V

vacancy Leerstelle
vacancy mechanism Leerstellendiffusion
vacancy sink Leerstellensenke
vacancy source Leerstellenquelle
vacuum etching Vakuumätzung; thermische Ätzung
valence band Valenzband
van der Waals bond Van-der-Waalssche Bindung
Vegard's law Vegardsche Regel
vermicular graphite vermikularer, verdichteter Graphit
vicinal plane fehlorientierte, vizinale Oberfläche
viscoelasticity Viskoelasizität
vitreous phase Glasphase
vitrification Glasbildung
Vollmer–Weber growth mode Vollmer-Weber-Wachstum[modus]
vol% Volumenanteil; Vol-%
volume diffusion Volumendiffusion

W

ω-phase ω-Phase
Wagner–Lifshitz–Slyozov theory Wagner-Lifshitz-Slyozov-Theorie
warm deformation Warmverformung
warm worked warmverformt
Warren–Averbach method Warren-Averbach-Verfahren
wavelength-dispersive spectrometry (WDS) wellenlängendispersive Spektrometrie
wavelength spectrum Wellenlängenspektrum
weak-beam imaging Weak-Beam-Dunkelfeldabbildung
wedge disclination Keildisklination
wedge fringes Keilinterferenzen; Dickenkonturen; Streifenkontrast
Weiss zone law Weissscher Zonengesetz
well-defined yield point ausgeprägte Streckgrenze

whisker Fadenkristall; Haarkristall
white [cast] iron weißes Gußeisen; Hartguß
white-heart malleable [cast] iron weißer Temperguß
white radiation Bremsstrahlung; polychromatische Röntgenstrahlung
Widmannstätten ferrite Widmannstättenscher Ferrit
Widmannstätten structure Widmannstättensches Gefüge
work hardening Kaltverfestigung; Umformverfestigung; Verfestigung
wt% Gewichtsprozent; Gew-%; Masseprozent; Masse-%
Wulff net Wulffsches Netz
wurzite Wurzit

X

x-ray absorption spectrum Röntgenabsorptionsspektrum
x-ray diffraction (XRD) Röntgenstrahlungsbeugung
x-ray diffraction line Röntgenbeugungslinie; Röntgenbeugungsmaximum
x-ray emission spectrum Röntgenemissionsspektrum
x-ray fluorescence Röntgenfluoreszenz
x-ray line intensity [Röntgen]linienintensität
x-ray line width Röntgenlinienbreite
x-ray microscopy Röntgenmikroskopie; Röntgentopographie
x-ray photoelectron spectroscopy (XPS) Röntgen-Photoelektronenspektroskopie
x-ray scattering Röntgenstrahlungsstreuung
x-ray spectroscopy Röntgenspektroskopie
x-ray structure analysis Röntgen-Phasenanalyse
x-ray topography Röntgentopographie; Röntgenmikroskopie

Y

yield point elongation Lüderssche Dehnung
yield stress Fließgrenze; Streckgrenze
yield strength Streckgrenze
Young's modulus Youngsches Modul; Elastizitätsmodul; E-Modul

Z

Zener drag Zenersche Hemmungskraft
Zener peak or relaxation Zener-Dämpfungsmaximum/-Relaxation
zinc blende [structure] type Zinkblende-Strukturtyp/-Typ
zirconia-toughened alumina (ZTA) zirkoniumoxid-verstärktes Aluminiumoxid
ZnS cubic structure Zinkblendegitter
zonal segregation Makroseigerung; Blockseigerung; Stückseigerung
zone Zone

zone annealing Zonenglühen
zone axis Zonenachse

German–English

60°-Versetzung 60°-dislocation

A

α-Eisen α-Fe
α-isomorphes Ti-X-System α isomorphous Ti–X system
α-Phase [in Ti-Legierungen] α-phase [in Ti alloys]
α′-Martensit α′-martensite
α″-Martensit α″-martensite
α-Titan α-Ti
α-Titanlegierung α Ti alloy
(α + β)-Messing (α + β) brass
(α + β)-Titanlegierung (α + β) Ti alloy
A_1–Ae_1-Temperatur A_1–Ae_1 temperature
A_2–Ae_2-Temperatur A_2–Ae_2 temperature
A_3–Ae_3-Temperatur A_3–Ae_3 temperature
A_4–Ae_4-Temperatur A_4–Ae_4 temperature
A_{cm}–Ae_{cm}-Temperatur A_{cm}–Ae_{cm} temperature
Abbildungsfehler aberration
Abbildungsmaßstab magnification
Abdruck replica
Aberration aberration
Abguß casting
Abschlußbereich closing domain
Abschlußbezirk closing domain
Abschreckalterung quench aging
Abschreckhärtung quench hardening
Abschreckung quenching
Absorption absorption
Absorptionsfaktor absorption factor
Absorptionskante absorption edge
Absorptionskoeffizient absorption or extinction coefficient
Absorptionskontrast absorption contrast
Absorptionsspektrum absorption spectrum
Achromat achromatic lens or objective
Achse der leichtesten Magnetisierung easy magnetization direction
Achsenverhältnis axial ratio
Achsenwinkel axial angle
Ac-Temperatur Ac temperature

Adatom adatom
adiabatische Näherung adiabatic approximation
Adsorbat adsorbate
Adsorbent adsorbent
adsorbiertes Atom adatom
Adsorption adsorption
Äquivalenz-Diagram equivalence diagram
Akkomodationsverformung accommodation strain
aktiviertes Gleitsystem active slip system
Aktivierungsanalyse activation analysis
Aktivierungsenergie activation energy
Aktivierungsenthalpie activation enthalpy
Akzeptor acceptor
aliovalenter Dotierstoff aliovalent solute or impurity
allgemeine Korngrenze general or random grain boundary
allotriomorpher Kristall grain boundary allotriomorph
allotrope Form/Modifikation allotropic form or modification
allotrope Umwandlung allotropic change
Allotropie allotropy
allseitiger Druck hydrostatic pressure
allseitig flächenzentriertes Gitter face-centered lattice
Alpha-Messing alpha brass
alphastabilisierender Zusatz α-stabilizer
Altern aging treatment
Alterung aging
Alterungshärtung age hardening; precipitation strengthening or hardening
Aluminiumoxid α-Al_2O_3
ambipolare Diffusion ambipolar diffusion
amorpher Festkörper glass; amorphous solid
Amplitudenkontrast amplitude contrast
analytische Durchstrahlungs-Elektronenmikroskopie analytical electron
      microscopy (AEM)
Andrade-Kriechen Andrade creep
Anelastizität anelasticity
angelassener Martensit tempered martensite
anisotrop anisotropic
Anlaßätzung staining
Anlassen tempering [treatment]
Anlassen von Stahl-Martensit tempering of steel martensite
Anlassen von Titan-Martensit tempering of titanium martensite
anormale Kornvergrößerung abnormal, exaggerated, or discontinuous grain
      growth; secondary recrystallization
anormale Röntgenstrahlungstransmission anomalous x-ray transmission
anormaler Perlit abnormal pearlite
anormales Kornwachstum abnormal, exaggerated, or discontinuous grain
      growth; secondary recrystallization

Anschliff metallographic section or sample
Antiferromagnetikum antiferromagnetic
antiferromagnetische Curie-Temperatur antiferromagnetic Curie point
Antiphasen-Domäne antiphase domain
Antiphasen-Grenze antiphase boundary
Antistrukturatom antistructural atom; antisite defect
Aperturblende aperture diaphragm
Apochromat apochromatic lens or objective
Arrhenius-Beziehung Arrhenius equation
Artefakt artifact
Ar-Temperatur Ar temperature
Asterismus asterism
Astigmatismus astigmatism
ASTM-Korngröße grain size number
athermische Umwandlung athermal transformation
Atomabstand interatomic spacing
Atomanordnung atomic structure
atomare Fehlstelle point defect
Atomformfaktor atomic scattering factor
Atomgewicht atomic mass
Atomgröße atomic size
Atomkonzentration, At.-% at%
Atommasse atomic mass
Atompackungsfaktor atomic packing factor
Atomradius atomic radius
Atomsondenspektroskopie atom probe field ion microscopy (APFIM)
Atomvolumen atomic volume
Ätzfigur etch figure
Ätzgrübchen etch pit
aufgelöster Stoff solute
aufgespaltete Versetzung extended dislocation
Auflichtmikroskop optical microscope
Auflösungsgrenze resolution limit
Auger-Elektron Auger electron
Auger-Elektronenspektroskopie (AES) Auger-electron spectroscopy (AES)
Auger-Rastermikrosonde scanning Auger-electron microscope (SAM)
Ausbreitungskugel Ewald or reflection sphere; sphere of reflection
Ausdehnung dilation
ausgeprägte Kornorientierung preferred grain orientation
ausgeprägte Streckgrenze well-defined or sharp yield point
Aushärten aging treatment
Aushärten von Ti-Legierungen aging treatment of Ti alloys
aushärtet thermally hardened
Aushärtung precipitation strengthening or hardening; age hardening; aging
Aushärtung [in Ti-Legierungen] aging [in Ti alloys]
Ausheilung recovery

Auslagern aging treatment
Auslöschung extinction
Auslöschungsregel extinction rule
Auslösungsvermögen resolving power
Ausscheidung precipitation; precipitate particle
Ausscheidungsbehandlung precipitation treatment
Ausscheidungshärtung age hardening; precipitation strengthening or hardening
Ausscheidungshemmung particle drag
Ausscheidungsphase precipitated phase
Austauschatom substitutional atom
Austauschmischkristall substitutional solid solution
Austauschwechselwirkung exchange interaction
Austenit austenite
Austenitformungshärten ausforming; low-temperature thermo-mechanical
    treatment
Austenitisierung austenitization
austenitischer Stahl austenitic steel
austenitisch-ferritischer Stahl austenitic-ferritic steel
austenitisch-martensitischer Stahl austenitic-martensitic steel
austenitstabilisierendes Element, Zusatz austenite-stabilizer
Austenitstabilisierung austenite stabilization
Austenit-Temperaturbereich austenitic range
Austenitverformen ausforming
Ausziehabdruck extraction replica
Autoradiographie autoradiography
Avogadro-Konstante Avogadro number
Avogadrosche Zahl Avogadro number
Avramische Gleichung Avrami equation

B

β-Aluminiumoxid β-Al_2O_3
β-Eisen β-Fe
β-eutektoides Ti-X-System β eutectoid Ti–X system
β-isomorphes Ti-X-System β isomorphous Ti–X system
β-Phase [in Ti-Legierungen] β-phase [in Ti alloys]
β-Titan β-Ti
β-Titanlegierung β Ti alloy
β_m-Phase [in Ti-Legierungen] β_m-phase [in Ti alloys]
B_s-Temperatur bainite start temperature (B_s)
Bainit bainite
bainitischer Stahl bainitic steel
bainitische Umwandlung bainitic transformation
Bainitisierung austempering
Bainit-Temperaturbereich/-Temperaturintervall bainitic range
Bambus-Gefüge bamboo structure

Bänderstruktur band structure
Bandlücke band gap
Basisebene basal plane
Basisgleitung basal slip
Basisstoff base
basiszentriertes Gitter base-centered or based lattice
Bauschinger-Effekt Bauschinger effect
Bedampfung shadowing
Beimengung impurity
Bergauf-Diffusion uphill diffusion
Bestandteil component
Bestrahlungsdefekte irradiation defects
betastabilisierender Zusatz, Element β-stabilizer
Beugungsdiagram diffractogram
Beugungskontrast diffraction contrast
Beugungspunkt diffraction spot
Beugungswinkel diffraction angle
Biegekontur bend or extinction contour
Bikristall bicrystal
Bildfeld field-of-view
bimetallisch bimetallic
bimodal bimodal
Bindungsenergie bond energy
Binodale binodal
Blechtextur sheet texture
Bloch-Wand Bloch wall
Block ingot
Blockmartensit lath or massive or packet or blocky martensite
Blockseigerung zonal or major segregation; macrosegregation; liquation
Boltzmann-Konstante/-Faktor Boltzmann constant
Bordoni-Dämpfungsmaximum/-Relaxation Bordoni peak or relaxation
Borrmannsches Effekt Borrmann effect
Bragg-Gleichung Bragg's law
Braggscher Winkel Bragg angle
Braggsche Reflexionsbedingung Bragg [diffraction] condition
Braggsches Reflex Bragg reflection
Bravais-Gitter Bravais lattice
Bremsspektrum continuous [x-ray] spectrum
Bremsstrahlung white or polychromatic radiation
Brennen firing
Bronze bronze
Burger-Orientierungsbeziehung Burger orientation relationship
Burgers-Umlauf Burgers circuit
Burgers-Vektor Burgers vector

C

χ-Carbid χ-carbide
CaF₂-Gitter CaF$_2$ structure
Carbid carbide
Cäsiumchlorid-Strukturtyp/-Typ CsCl structure [type]
charakteristische Röntgenstrahlung characteristic x-rays
chemische Adsorption chemisorption
chemische Ätzung chemical etching
chemische Diffusion chemical diffusion; interdiffusion
chemische Inhomogenität chemical inhomogeneity
chemisches Potential chemical potential
cholesterischer Kristall N* crystal; cholesteric crystal
chromatische Aberration chromatic aberration
C-Kurve C-curve
Coble-Kriechen Coble creep
Compton-Streuung Compton scattering
Cottrell-Wolke Cottrell atmosphere or cloud
Cristobalit cristobalite
Crowdion crowdion
CsCl-Gitter CsCl structure
Curie-Temperatur (T$_C$, Θ$_C$) Curie point or temperature (T$_C$, Θ$_C$)

D

δ-Eisen δ-Fe
δ-Ferrit δ-ferrite
ΔR-Wert Δr-value
Dämpfung internal friction
Debye-Scherrer-Verfahren Debye–Scherrer method
Deformationsband deformation or transition band; microband
Dehnung tensile strain
dekorierte Versetzung decorated dislocation
Dendrit dendrite
Dendritseigerung coring; core or dendritic segregation
Desorption desorption
Diamagnetikum diamagnetic
Diamantgitter diamond structure
dichtgepackte Atomreihe close-packed direction or row
Dickenkonturen wedge or thickness fringes
Differenzial-Interferenzkontrast differential interference contrast
Differenzial-Rasterkalorimetrie differential scanning calorimetry (DSC)
Differentialthermoanalyse differential thermal analysis (DTA)
Diffraktionswinkel diffraction angle
Diffraktometer diffractometer
Diffraktometerverfahren diffractometric method

diffuse Streuung diffuse scattering
Diffusion diffusion
diffusionsabhängige Umwandlung diffusional transformation
diffusionsbestimmte diffusion-controlled
Diffusionsglühen homogenizing [anneal] or homogenization
diffusionsinduzierte Korngrenzenwanderung diffusion-induced grain boundary migration (DIGM)
diffusionsinduzierte Rekristallisation diffusion-induced recrystallization (DIR)
Diffusionskoeffizient diffusion coefficient; diffusivity
Diffusionskonstante diffusion coefficient; diffusivity
Diffusionskriechen diffusional creep
Diffusionsplastizität diffusional plasticity
Dilatometer dilatometer
Disklination disclination
diskontinuierliche Auflösung discontinuous dissolution
diskontinuierliche Ausscheidung discontinuous or cellular precipitation
diskontinuierliche Kornvergrößerung discontinuous, exaggerated, or abnormal grain growth; secondary recrystallization
diskontinuierliche plastische Verformung discontinuous yielding
diskontinuierliche Rekristallisation discontinuous recrystallization
diskontinuierliches Kornwachstum discontinuous, abnormal, or exaggerated grain growth; secondary recrystallization
diskontinuierliche Vergröberung discontinuous coarsening
Dispersionshärtung dispersion strengthening
Dispersionsphase dispersed phase
Dispersionsverfestigung dispersion strengthening
Dispersoid dispersoid
dispersoidfreie Zone dispersoid-free zone
Dodekaederebene dodecahedral plane
Domänenanordnung domain structure
Domänengefüge domain structure
Domänengrenze domain wall
Domänenstruktur magnetic structure
Domänenwand domain wall
Donator donor
Doppelaltern double aging
Doppelleerstelle divacancy
Doppelquergleitung double cross-slip
Doppelstapelfehler double or extrinsic stacking fault
Doppelversetzungskinke double kink
Dotierstoff dopant
Dotierung microalloying; doping
Drehkorngrenze twist grain boundary
Drehkristallverfahren rotating crystal method
dreistoff ternary

Dressieren temper rolling
Drift-Bewegung drift
DSC-Gitter displacement shift complete (DSC) lattice
Dublette doublet
Duktilitätsübergangstemperatur ductility transition [temperature]; ductile-
    brittle transition [temperature]
duktil-spröde Übergangstemperatur ductile-brittle transition [temperature];
    ductility transition [temperature]
Dunkelfeldabbildung dark-field image
Dunkelfeld-Beleuchtung dark-field illumination
Dunkelfeldbild dark-field image
Duplexgefüge duplex microstructure
Duplexkorngröße duplex grain size
durchschnittliche Größe mean [grain or particle] size
Durchstrahl-Elektronenmikroskop transmission electron microscope (TEM)
Durchwärmen soaking
dynamische Erholung dynamic recovery
dynamische Reckalterung Portevin–Le Chatelier effect; dynamic strain aging
dynamische Rekristallisation dynamic recrystallization

E

ε-Karbid ε-carbide
ε-Martensit ε-martensite
η-Karbid η-carbide
Easy Glide easy glide
eigenes Zwischengitteratom self-interstitial; interstitialcy
Eigenspannung 1. Art macroscopic stress
Eigenspannung 2. Art microscopic stress
Eigenspannung 3. Art static lattice distortion
Eigenspannungen internal or residual stresses
Eindomänen-Teilchen single-domain particle
einfache Gleitung single slip
einfaches Gitter, Raumgitter simple or primitive lattice
eingeschreckte Leerstelle quench-in vacancy
Einheitszelle unit cell
Einkristall single crystal
einkristallin monocrystalline
Einlagerungsmischkristall interstitial solid solution
Einlagerungsphase interstitial phase or compound
einseitig flächenzentriertes Gitter base-centered lattice
einstufiger Abdruck direct replica
Einweg-Formgedächtniseffekt one-way shape memory effect
elastische Streuung elastic or coherent scattering
elastische Verformung elastic deformation
Elastizitätsenergie elastic strain energy

Elastizitätsmodul Young's or elastic modulus; modulus of elasticity
elektrolytische Ätzung electro-etching
Elektromigration electromigration
Elektronegativität electronegativity
Elektronenbeugung electron diffraction
Elektronenbeugungsbild electron diffraction pattern
Elektronenenergie-Verlustspektroskopie electron energy loss spectroscopy
    (EELS)
Elektronenkanalierung electron channeling
Elektronen-Kanalierungsbild electron channeling pattern (ECP)
Elektronenkonzentration electron concentration
Elektronenmikroskopie electron microscopy (EM)
elektronenmikroskopische Aufnahme/Abbildung electron micrograph
Elektronenmikrosonden-Analyse electron probe microanalysis (EPMA)
Elektronenspektroskopie zur chemischen Analyse electron spectroscopy for
    chemical analysis (ESCA)
Elektronenstrahl-Mikrosonde electron [micro]probe; microprobe
elektronische Verbindung/Phase Hume-Rothery phase; electron compound or
    phase
Elektrotransport electromigration
elementarer Versetzungssprung elementary jog
Elementarzelle unit cell
Elementarzellenparameter unit cell parameter
E-Modul Young's or elastic modulus; modulus of elasticity
Empfindlichkeit zur Formänderungsgeschwindigkeit strain rate sensitivity
Endmischkristall terminal or primary solid solution
energiedispersive Röntgenanalyse energy-dispersive diffractometry
    (EDS/EDAX)
energiedispersive Röntgenspektroskopie energy-dispersive spectrometry
Energielücke band gap; forbidden gap
Energiespektrum energy spectrum
entarteter Eutektoid divorced eutectoid
entarteter Perlit divorced pearlite
Entglasen devitrification
Entmischung decomposition
Entmischungszone Guinier–Preston (GP) zone
Entspannung stress relaxation
Entspannungsglühen stress-relief anneal
epitaktische Dünnschicht epitaxial film
epitaktische Versetzung epitaxial dislocation
Epitaxie epitaxy
Equilibriumphase equilibrium or stable phase
Erholung recovery
Erstarrung solidification; crystallization
Erstarrungspunkt solidification point or temperature
Erstarrungstemperatur solidification point or temperature

Eulersche Winkel Euler angles
Eutektikum eutectic [structure]
eutektische Kolonie eutectic colony
eutektische Reaktion eutectic reaction
eutektischer Punkt eutectic point
eutektisches Gemisch eutectic [structure]
eutektisches Korn eutectic colony or grain
eutektische Temperatur eutectic temperature
Eutektoid eutectoid [structure]
eutektoide Entmischung eutectoid decomposition
eutektoide Kolonie eutectoid colony
eutektoide Reaktion eutectoid reaction
eutektoider Punkt eutectoid point
eutektoider Zerfall eutectoid decomposition
eutektoide Temperatur eutectoid temperature
Ewaldsche Kugel Ewald or reflection sphere; sphere of reflection
exponentielles Kriechen power-law creep
Extinktion extinction
Extinktionskontur bend or extinction contour
Extraktionsabdruck extraction replica
extrinsische Korngrenzenversetzung extrinsic grain-boundary dislocation
extrinsischer Stapelfehler extrinsic stacking fault

F

Fadenkristall whisker
Farbätzung tinting; color etching
Farbniederschlagsätzung tinting; color etching
Farbzentrum color center
Fasergefüge banded structure
Fasertextur fiber texture
fast-spezielle Korngrenze nearly special grain boundary
Fe–C System Fe–C system
Fe–Fe₃C System Fe–Fe₃C system
fehlgeordneter Mischkristall disordered solid solution
Fehlordnung disordering
fehlorientierte Oberfläche vicinal plane
Fehlpassungsparameter misfit parameter
Fehlpassungsversetzung misfit dislocation
Fehlstruktur defect structure or lattice
Feinbereichsbeugung microdiffraction; selected area diffraction (SAD/ESAD)
Feinbereichskanalierungsbild selected area channeling pattern (SACP)
Feingefüge fine or dislocation structure; substructure
Feinkornhärtung grain-boundary strengthening
feinkörnig fine-grained
feinkristallin fine-grained

Feldemission field or autoelectronic emission
Feldionenmikroskop (FIM) field-ion microscope (FIM)
ferngeordneter Mischkristall ordered solid solution
Fernordnung long-range order
Fernordnungsgrad degree of long-range order
Fernordnungsparameter long-range order parameter
Ferrimagnetikum ferrimagnetic
Ferrit ferrite
Ferrit-Granat garnet ferrite
ferritischer Stahl ferritic steel
ferritisches Gußeisen ferritic [cast] iron
Ferrit-Spinell spinel ferrite
ferritstabilisierendes Element ferrite-stabilizer
Ferroelektrikum ferroelectric
ferroelektrische Domäne ferroelectric domain
Ferromagnetikum ferromagnetic
fest solid
feste Lösung solid solution
Festigkeitseigenschaft mechanical property
Festphasensintern solid-state sintering
1. Ficksches Gesetz Fick's first law
2. Ficksches Gesetz Fick's second law
Filmabdruck direct replica
flache Störungsstelle shallow impurity
Flächenanisotropie planar anisotropy
flächenförmiger Gitterfehler planar defect
Flächenhäufigkeitsfaktor multiplicity factor
Fließfigur stretcher-strain marking
Fließgrenze yield stress
Fließkurve stress-strain diagram
Fließspannung flow stress
Flußspatgitter CaF_2 structure
Fluktuation fluctuation
Fluorit fluorite
Fluoritgitter CaF_2 structure
Fluorit-Strukturtyp/-Typ fluorite [structure] type
Flüssigkristall liquid crystal
Flüssigphasensintern liquid-phase sintering
Folie thin foil
Form form
Formänderung strain
Formänderungsgeschwindigkeit strain rate
Formgedächtniseffekt shape memory effect
Fotoemissions-Elektronenmikroskop photo-electron emission microscope
(PEEM)
Fragmentierung fragmentation

Frank–Read-Quelle Frank–Read source
Frank–van-der-Merve-Wachstum[modus] Frank–van der Merve growth mode
Frank-Vektor Frank vector
Franksche Teilversetzung Frank partial dislocation
freie Energie Helmholtz free energy
freie Enthalpie Gibbs' free energy; free enthalpy
freies Exzeßvolumen, Volumen excess free volume
Freiheitsgrad degree of freedom
Fremdatom foreign atom
Fremdatomendiffusion solute diffusion
Fremdatomenhemmung* solute drag
Fremdatomenwolke impurity cloud
fremdes Zwischengitteratom interstitial [foreign] atom; interstitial
Frenkel-Paar Frenkel pair
Furchenhemmung* groove drag
F-Zentrum F-center

G

γ-Eisen γ-Fe
γ'-Phase γ'-phase
Gaskonstante gas constant
Gefrierpunkt solidification point or temperature
Gefüge structure
gefüge-abhängig structure-sensitive
Gefügebestandteil microconstituent
Gefügenelement microconstituent
gefüge-unabhängig structure-insensitive
gekoppeltes Wachstum cooperative growth
gekoppelte Umwandlung nondiffusional transformation
gekoppeltes Wachstum coupled growth
gekoppelte Umwandlung diffusionless transformation
gemischte Korngrenze mixed grain boundary
gemischte Versetzung mixed dislocation
geometrische Koaleszenz geometric coalescence
geometrisch-notwendige Versetzung* geometrically necessary dislocations
gerichtete Erstarrung directional solidification
gesättigter Mischkristall saturated solid solution
gespeicherte Verformungsenergie stored energy
Gewichtsprozent, Gew.-% wt%
Gibbssche freie Energie Gibbs' free energy
Gibbssches Phasengesetz Gibbs' phase law
Gibbssche Phasenregel Gibbs' phase rule
Gibbs–Thomson-Gleichung Gibbs–Thomson equation
Gire symmetry axis
Gitter lattice

gitter-anpassende [epitaktische] Dünnschicht* lattice-matched [epitaxial] film
Gitterbasis lattice basis
Gitterbaufehler crystal defect or imperfection
Gitterebene lattice plane
Gitterebene hoher Belegungsdichte densely packed plane
Gitterebene maximaler Belegungsdichte close-packed plane
gitter-fehlpassende [epitaktische] Dünnschicht * lattice-mismatched [epitaxial] film
Gitterfehlpassung lattice misfit
Gittergerade lattice direction
Gittergerade dichtester Besetzung close-packed direction or row
Gittergrundvektor fundamental translation vector
Gitterhohlraum lattice void
Gitterknoten lattice point or site
Gitterkonstante lattice constant
Gitterlücke interstice; lattice void
Gitterparameter lattice parameter
Gitterpperiode lattice constant
Gitterpunkt lattice point or site
Gitterpunktabstand interatomic spacing
Glanzwinkel glancing or Bragg angle
glasartiger Festkörper, Substanz amorphous solid
Glasbildung vitrification
Glaskeramik glass-ceramic
Glasphase glassy or vitreous phase
Glasübergangstemperatur (T_g) glass transition temperature (T_g)
gleichachsig equiaxed
Gleichgewichtslinie transus
Gleichgewichtsphase equilibrium or stable phase
Gleichgewichtsschaubild equilibrium, phase, or constitution diagram
Gleichgewichtssegregation equilibrium segregation
Gleichgewichtssystem equilibrium system
Gleichgewichtstemperatur equilibrium temperature
Gleitband slip band
Gleitebene slip plane
gleitfähig glissile
Gleitlinie slip line
Gleitmodul shear modulus
Gleitrichtung slip direction
Gleitspur slip trace
Gleitsystem slip system
Gleitung slip
Gleitvorgang slip; glide
globularer Perlit spheroidite; spheroidized pearlite
Glühen annealing or anneal

Glühtextur annealing texture
Glühzwilling annealing twin
Goss-Lage/-Textur Goss or cube-on-edge texture
Gräbchenhemmung* groove drag
Graphit graphite
Graphitisierungsglühen graphitization [annealing]
graphitstabilisierendes Element, Zusatz graphitizer
graues Gußeisen gray [cast] iron
Grauguß gray [cast] iron
Greninger-Troiano-Orientierungsbeziehung Greninger–Troiano orientation
      relationship
Grenzfläche interface
Größenverteilung size distribution
grober Perlit coarse pearlite
Grobgefüge macrostructure
grobkörnig coarse-grained
Grobkornrekristallisation discontinuous, abnormal, or exaggerated grain
      growth; secondary recrystallization
Großwinkelkorngrenze high-angle or large-angle grain boundary
Grunddreieck standard stereographic triangle
Grundmasse matrix
Grundstoff base
Gußeisen cast iron
Gußeisen mit Knotengraphit/Kugelgraphit nodular or ductile [cast] iron
Gußeisen mit Lamellengraphit gray [cast] iron
Gußstück casting
Guinier-Preston-Zone Guinier–Preston (GP) zone

H

Haarkristall whisker
Habitus habit
Habitusebene habit plane
Halbwertsbreite full width at half maximum (FWHM)
Hall-Petch-Gleichung Hall–Petch equation
Haltepunkt arrest or critical point
Harper-Dorn-Kriechen Harper–Dorn creep
Härtbarkeit hardenability
Härte hardness
Härten hardening [treatment]; quenching
Hartguß white [cast] iron
Hauptgleitsystem active or primary slip system
hdp Struktur hexagonal close-packed (HCP) structure
Hebelbeziehung lever rule
Hebellinie tie line; conode
heiß-isostatisches Pressen hot isostatic pressing (HIP)

Heißpressen hot pressing
Heiztischmikroskop hot-stage microscope
Hellfeldabbildung bright-field image
Hellfeld-Beleuchtung bright-field illumination
Hellfeldbild bright-field image
Helmholtzsche freie Energie free energy
hemmende Kraft pinning or drag force
Hemmungskraft pinning or drag force
heteroepitaktische Dünnschicht heteroepitaxial film
heterogene Keimbildung heterogeneous nucleation
heterogenes Gefüge heterogeneous microstructure
heterogenes System heterogeneous system
heteropolare Bindung heteropolar or ionic bond
Heterostruktur heterostructure
Heteroübergang heterojunction
Hexaferrit hexagonal ferrite
hexagonal-dichtestgepackte Struktur hexagonal close-packed (HCP) structure
hexagonales System hexagonal system
Hochauflösungs-Elektronenmikroskop high-resolution transmission electron
    microscope (HRTEM)
Hochspannungs-Elektronenmikroskop high-voltage electron microscope
    (HVEM)
Hochtemperaturmikroskop hot-stage microscope
hochtemperatur thermomechanische Behandlung high-temperature thermo-
    mechanical treatment
homoepitaktische Dünnschicht homoepitaxial film
homogene Keimbildung homogeneous nucleation
homogenes Gefüge homogeneous microstructure
homogenes System homogeneous system
Homogenisierung homogenizing [anneal]; homogenization
homologische Temperatur homologous temperature
homopolare Bindung homopolar or covalent bond
Hookesches Gesetz Hooke's law
Hume–Rothery-Phase Hume–Rothery phase; electron compound or phase
Hume–Rothery-Regeln Hume–Rothery rules
hydrostatischer Druck hydrostatic pressure

I

Ideallage ideal orientation; texture component
Immersionsobjektiv immersion objective or lens
Impfkristall seed crystal
Impfstoff inoculant
inhomogenes Gefüge inhomogeneous microstructure
inkohärente Grenzfläche incoherent interface
inkohärentes Ausscheidungsteilchen incoherent precipitate or particle

inkohärente Zwillingsgrenze incoherent twin boundary
Inkubationszeit incubation or induction period
innere Oxidation internal oxidation
innere Reibung internal friction
innere Spannungen internal stresses
Inseldünnschicht island film
in situ-**Beobachtung** *in situ* observation
in situ-**Rekristallisation** continuous recrystallization; recrystallization *in situ*
instrumentale Linienverbreiterung instrumental [x-ray] line broadening
Integrallinienbreite integral [x-ray] line width
Integral[linien]intensität integrated [x-ray] line intensity
Interdiffusion interdiffusion; chemical diffusion
Interferenzdiagram diffractogram
Interferenzschliere bend or extinction contour
interkritischer Umwandlungsbereich intercritical range
interkritische Wärmebehandlung intercritical heat treatment
interkristallin intercrystalline
intermediäre Phase intermediate phase
intermetallische Verbindung intermetallic compound
Interphasenausscheidung interphase precipitation
interstitielle Phase interstitial compound or phase
interstitieller Diffusionsmechanismus interstitial [mechanism of] diffusion
interstitieller Mischkristall interstitial solid solution
interstitielles Fremdatom interstitial [foreign] atom; interstitial
intrakristallin intracrystalline; intragranular; transcrystallin; transgranular
intrinsische Diffusionskonstante intrinsic diffusion coefficient or diffusivity
intrinsische Korngrenzenversetzung intrinsic grain-boundary dislocation
intrinsischer Stapelfehler intrinsic stacking fault
invariante Reaktion invariant reaction
Ionenätzung ion etching
Ionenbindung ionic or heteropolar bond
Ionenimplantation ion implantation
Ionenkanalierung ion channeling
Ionenkristall ionic crystal
Ionenradius ionic radius
irreversibel irreversible
isochrones Glühen isochronal annealing
Isoforming isoforming
isomorphe Phasen isomorphous phases
isomorphes System isomorphous system
Isomorphie isomorphism
isotherm isothermal
isothermes Härten marquenching; martempering
isotherme Umwandlung isothermal transformation
isothermes Zeit-Temperatur-Umwandlung-ZTU-Schaubild isothermal transformation (TTT) diagram

isothermes ZTU-Diagram/-Schaubild isothermal transformation (TTT) dia-
     gram
isotrop isotropic

J

Johnson–Mehl–Kolmogorov-Gleichung Johnson–Mehl–Kolmogorov equation

K

Kaloriemetrie calorimetry
kaloriemetrische Analyse calorimetry
Kaltaushärtung natural aging
Kaltauslagerung natural aging
Kaltverfestigung work or strain hardening
kaltverformt cold worked
Kaltverformung cold deformation
Kantenversetzung stair-rod dislocation
kapillare Triebkraft* capillary driving force
Karbid carbide
Karbidbildner carbide-former
Karbidseigerung carbide segregation
Karbidzellgefüge carbide network
Karbonitrid carbonitride
Kê-Dämfungsmaximum/-Relaxation Kê peak or relaxation
Keildisklination wedge disclination
Keilinterferenzen wedge or thickness fringes
Keim nucleus
Keimbildner nucleation agent
Keimbildung nucleation
Keimbildungsgeschwindigkeit nucleation rate
Kelvin-Tetrakaidekaeder Kelvin's tetrakaidecahedron
Kerr-Mikroskopie Kerr microscopy
kfz Struktur face-centered cubic (FCC) structure
Kikuchi-Linien Kikuchi lines
Kippgrenze tilt grain boundary
Kirkendall-Effekt Kirkendall effect
Kirkendall-Porosität diffusion porosity
Kleinwinkelkorngrenze subboundary; subgrain boundary; low-angle or small-
     angle grain boundary
Knickband kink band
Knötchengraphit nodular or spheroidal graphite
Koagulation coagulation
Koaleszenz coalescence
kohärente Grenzfläche coherent interface

kohärentes Ausscheidungsteilchen coherent precipitate
kohärente Zwillingsgrenze coherent twin boundary
Kohärenzspannung accommodation or coherency strain
Kohärenzspannungs-Verfestigung* coherency strain hardening
Kohlenstoffstahl plain carbon steel
Koinzidenzgitter coincidence site lattice
Koinzidenzgrenze special grain boundary; CSL-boundary
Kolonie colony
Kompatibilitäts-Diagram compatibility diagram
Kompensationsokular compensating eyepiece
Komplexkarbid complex carbide
Komponente component
Kompressionsmodul bulk modulus
Kompromißtextur compromise texture
kondensierte Wolke condensed atmosphere
kongruent congruent
konjugiertes Gleitsystem conjugate slip system
Konode conode; tie line
Konodenregel lever rule
konstitutionelle Unterkühlung constitutional undercooling or supercooling
kontinuierliche Ausscheidung, Entmischung continuous precipitation
kontinuierliche Kornvergrößerung continuous or normal grain growth
kontinuierliche Rekristallisation continuous recrystallization; recrystallization
    in situ
kontinuierlicher Zerfall continuous precipitation
kontinuierliches Kornwachstum continuous or normal grain growth
kontinuierliches [Röntgen]spektrum continuous [x-ray] spectrum
kontrolliertes Walzen controlled rolling
konvergente Elektronenbeugung convergent beam electron diffraction
    (CBED)
Konzentration concentration
Koordinationspolyeder coordination polyhedron
Koordinationssphäre coordination shell
Koordinationszahl coordination number
Korn grain; crystallite
Kornfeinung grain refining
Korngrenze grain boundary
Korngrenzenbeweglichkeit grain-boundary mobility
Korngrenzencharakter-Verteilung* grain-boundary character distribution
Korngrenzendiffusion grain-boundary diffusion
Korngrenzendrehmoment grain-boundary torque
Korngrenzenenergie grain-boundary energy
Korngrenzenflächenspannung grain-boundary energy or tension
Korngrenzengleitung grain-boundary sliding
Korngrenzenhärtung grain-boundary strengthening
Korngrenzenlage grain-boundary orientation

Korngrenzenmobilität grain-boundary mobility
Korngrenzensegregation grain-boundary segregation
Korngrenzenversetzung grain boundary dislocation
Korngröße grain size
Korngrößenhomogenität grain size homogeneity
Korngrößenkennzahl grain size number
körniger Bainit upper bainite
körniger Perlit granular or spheroidized pearlite; spheroidite
Kornkante triple joint
Kornkoaleszenz grain coalescence
kornorientiert grain-oriented; textured
Kornseigerung coring; core or dendritic segregation; microsegregation
Kornstreckung grain aspect ratio
Kornvergröberung grain coarsening
Kornwachstum grain growth
Kornwachstum induziert durch Korngrenzenkrümmung* curvature-driven
      grain growth
Kornwachstumsgeschwindigkeit grain growth rate
Korund corundum
Kossel-Linienbild Kossel line pattern
Köster-Effekt Köster effect
kovalente Bindung covalent or homopolar bond
kovalenter Radius covalent radius
kovalentes Kristall covalent crystal
Kraftwirkungsfigur stretcher-strain marking
Kriechen creep
Kriechkavitation creep cavitation
Kristall crystal
Kristallachse crystal axis
Kristallbaufehler crystal defect or imperfection
Kristallfigur etch figure
Kristallgitter point or crystal lattice
kristallin crystalline
kristalline Anisotropie crystalline anisotropy
kristalline Keramik crystalline ceramic
kristalliner Bruch crystalline fracture
Kristallisation crystallization
Kristallisationspunkt crystallization point or temperature
Kristallisationstemperatur crystallization point or temperature
Kristallit grain; crystallite
Kristallkorn grain; crystallite
Kristallmonochromator crystal monochromator
kristallographische Textur texture
Kristallseigerung coring; core or dendritic segregation; microsegregation
Kristallstruktur crystal structure
Kristallsystem crystal system

kritische Abkühlungsgeschwindigkeit critical cooling rate
kritische resultierende Schubspannung critical resolved shear stress
kritischer Keim critical [size] nucleus
kritischer Verformungsgrad/Reckgrad critical deformation
krz Struktur body-centered cubic (BCC) structure
kubischer Martensit cubic martensite
kubisches System cubic system
kubisch-flächenzentrierte Struktur face-centerd cubic (FCC) structure
kubisch-raumzentrierte Struktur body-centered cubic (BCC) structure
Kugelbildung spheroidization
Kugelgraphit spheroidal or nodular graphite
kugeliger Perlit spheroidite
Kupfer-Lage Cu-type orientation
Kurdjumov-Sachs-Orientierungsbeziehung Kurdjumov–Sachs orientation
    relationship
kurz-reichend short-range

L

Lagekugel reference or orientation sphere
Lamellenabstand interlamellar spacing
Lamellengraphit flake graphite
laminare Gleitung laminar slip
Lankford-Wert Lankford coefficient; $\bar{r}$-value
lanzettförmiger Martensit lath, massive, packet, or blocky martensite
Larson-Miller-Parameter Larson–Miller parameter
latente Verfestigung latent hardening
Lattenmartensit acicular, plate, or lenticular martensite
Laue-Aufnahme Laue diffraction pattern
Laue-Bedingungen/-Gleichungen Laue equations
Lauemethode Laue method
Laue-Reflexionsmethode back-reflection Laue method
Laue-Reflexionsverfahren back-reflection Laue method
Laue-Transmissionsverfahren transmission Laue method
Laueverfahren Laue method
Laves-Phase Laves phase
Ledeburit ledeburite
ledeburitischer Stahl ledeburitic steel
Leerstelle vacancy
Leerstellendiffusion vacancy mechanism
Leerstellenquelle vacancy source
Leerstellensenke vacancy sink
Legierung alloy
Legierungselement alloying element
Legierungssystem alloy system
Leitungsband conduction band

Leuchtfeldblende field diaphragm
Lichtmikroskop optical microscope
linearer Defekt linear defect
linearer Schwächungskoeffizient linear absorption coefficient
lineare Wachstumsgeschwindigkeit linear growth rate
linienförmiger Gitterfehler linear defect
Linienverbreiterung line broadening
linsenförmiger Martensit lenticular, acicular, or plate martensite
Liquidus[kurve] liquidus
Liquidus[linie] liquidus
logarithmisches Kriechen logarithmic creep
Lomer-Cottrell-Schwelle Lomer–Cottrell barrier or lock
Lorentz-Faktor Lorentz factor
Lorentz-Mikroskopie Lorentz microscopy
Lösligkeit im festen Zustand solid solubility
Lösligkeitsgrenze solubility limit
Lösligkeitslinie solvus
Lösungsbehandlung solution treatment
Lüderssche Dehnung Lüders strain; yield point elongation
Lüdersscher Band Lüders band
Luftabkühlung air-cooling

M

M_d-Temperatur M_d temperature
M_s^σ-Temperatur M_s^σ temperature
magnetische Bereichsgefüge/Bereichsstruktur magnetic or domain structure
magnetische Kristallanisotropie magnetic crystalline anisotropy
magnetische Ordnung magnetic ordering
magnetischer Bezirk magnetic domain
magnetische Textur magnetic texture
magnetische Unwandlung magnetic transformation
Magnetkraft-Mikroskop magnetic force microscope (MFM)
Makroaufnahme macrograph
Makroeigenspannung macroscopic stress
Makrogefüge macrostructure
Makroseigerung macrosegregation; major or zonal segregation; liquation
Martensit martensite
martensitaushärtender Stahl maraging steel
martensitischer Stahl martensitic steel
Martensit-Temperaturbereich/-Temperaturintervall martensitic range
Martensitumwandlung martensitic transformation
Massenschwächungskoeffizient mass absorption coefficient
Massenzahl atomic mass
Masseprozent, Masse-% wt%
massiver Martensit massive, block, lath, or packet martensite

massive Umwandlung massive transformation
Matrix matrix
Matrixatom host atom
Matrixband matrix band
Matrixstoff solvent
Matrizenabdruck indirect replica
Matthiessen-Regel Matthiessen's rule
mechanische Anisotropie mechanical anisotropy
mechanische Eigenschaft mechanical property
mechanisches Legieren mechanical alloying
mechanische Stabilisation [von Austenit] mechanical stabilization [of austen-
      ite]
mechanische Zwillingsbildung deformation twinning
Mediangröße median size
mehrfache Gleitung multiple slip
mehrfache Quergleitung multiple cross slip
mehrfacher Versetzungssprung multiple jog
mehrphasig heterophase
Mesophase mesomorphic phase; mesophase
Messing brass
Messing-Lage Bs/Def orientation
Messing-Rekristallisationslage Bs/Rex orientation
metadynamische Rekristallisation metadynamic recrystallization
Metallbindung metallic bond
Metallid intermetallic compound
metallischer Radius metallic radius
metallisches Glas metallic glass
metallisches Kristall metallic crystal
Metallkeramik metal ceramic
Metallmikroskop optical microscope
metallographische Untersuchung metallographic examination
[metallographischer] Schliff metallographic section or sample
metastabile β-Legierung metastable β alloy
metastabile β-Phase (β_m) metastable β-phase (β_m)
metastabile Phase metastable phase
metastabiler Zustand metastable state
Mikroanalyse microanalysis
Mikroaufnahme micrograph
Mikroband microband; transition or deformation band
Mikrodehnung microstrain
Mikroeigenspannung microscopic stress
Mikrogefüge microstructure
Mikrolegieren microalloying; doping
Mikroseigerung coring; core or dendritic segregation
Mikrosonde microprobe
Mikrostruktur microstructure

Mikrotextur microtexture
Mikroverformung microstrain
Miller-Bravais-Indizes Miller–Bravais indices
Millersche Indizes Miller indices
Mischkarbid complex or alloy carbide
Mischkristall solid solution
Mischkristallhärtung solid solution strengthening or hardening
Mischkristallverfestigung solid solution strengthening or hardening
Mischkristallzersetzung decomposition
Mischungslücke miscibility gap
Misorientation misorientation; disorientation
Misorientations-Verteilungsfunktion misorientation distribution function
Mittelrippe midrib
mittlere Größe mean size
Modalgröße most probable size; mode
Modifizierung modification
moduliertes Gefüge modulated structure
Modus mode; most probable size
Moiré-Bild moiré pattern
Mole-% mol%
monochromatische Strahlung monochromatic radiation
monoklines System monoclinic system
monotektische Reaktion monotectic reaction
monotektoide Reaktion monotectoid reaction
Mosaikgefüge mosaic structure
M-Zentrum M-center

N

Nabarro-Herring-Kriechen Nabarro–Herring creep
Nachwalzen temper rolling
Nachwirkung aftereffect
nadelförmig acicular
nadelförmiger Ferrit acicular ferrite
nadelförmiger Martensit acicular or plate martensite
nadeliger Bainit lower bainite
Nahordnung short-range order
Nahordnungsparameter short-range order parameter
Nahordnungszustand short-range order
nanokristallin nanocrystalline
nanostrukturell nanocrystalline
Natriumchloridgitter NaCl structure
Néel-Temperatur (T_N, Θ_N) Néel point/temperature (T_N, Θ_N); antiferromagnetic
     Curie point
Néel-Wand Néel wall
nematischer Kristall nematic crystal; N crystal

Nenndehnung engineering strain
Nennspannung nominal stress
Netzebene net plane; lattice plane
Netzebenenabstand interplanar spacing
Neumannscher Band Neumann band
Neutronenbeugung neutron diffraction
nichtgesättigter Mischkristall unsaturated [solid] solution
nicht-gleitfähige Versetzung sessile dislocation
Niedrigenergetischenelektronenbeugung low-energy electron diffraction (LEED)
Nishiyama-Wassermann-Orientierungsbeziehung Nishiyama orientation relationship
Nitrid nitride
normales Kornwachstum normal or continuous grain growth
normale Spannung normal stress
Normalglühen normalizing
Normalisieren normalizing
numerische Apertur numerical aperture
n-zählige Drehachse n-fold axis

O

ω-Phase ω-phase
obere Fließgrenze upper yield stress
oberer Bainit upper bainite
Oberflächenenergie-Triebkraft* surface-energy driving force
Oberflächenfurche thermal groove
Oberflächenspannung surface tension
Öffnungsfehler spherical aberration
Oktaederebene octahedral plane
Oktaedergitterlücke octahedral interstice, site, or void
Oktaederhohlraum octahedral interstice, site, or void
Oktaeder-Zwischengitterplatz octahedral interstice, site, or void
Orangenhaut orange peel
Ordnungsumwandlung order–disorder transformation or transition; long-range ordering
Ordnungs-Unordnungs-Umwandlung order–disorder transformation or transition; long-range ordering
orientierungsabbildende Mikroskopie orientation imaging microscopy (OIM)
Orientierungsbeziehung orientation relationship
Orientierungskontrast diffraction contrast
Orientierungsverteilungsfunktion (OVF) orientation distribution function (ODF)
Orientierungszusammenhang orientation relationship
Orowan-Mechanismus Orowan mechanism
Orowan-Schleife Orowan loop

Orthoferrit orthoferrite
orthorhombisches System orthorhombic system
Ostwald-Reifung Ostwald ripening
oxidendispersionsgehärtet oxide dispersion strengthened (ODS)
Oxynitrid oxynitride

P

Packungsdichte packing factor
Packungsfaktor packing factor
Paramagnetikum paramagnetic
Partialversetzung partial or imperfect dislocation
Partikelverstärkung dispersion strengthening
Peierls-Spannung/-Schwelle Peierls stress/barrier
Pencile Glide pencil glide
peritektische Reaktion peritectic reaction
peritektische Temperatur peritectic temperature
peritektoide Reaktion peritectoid reaction
Perlit pearlite
Perlitbildung pearlitic transformation
perlitischer Gußeisen pearlitic [cast] iron
perlitischer Stahl pearlitic steel
perlitische Umwandlung pearlitic transformation
Perlitkolonie pearlitic colony or nodule
Perlitkorn pearlitic colony or nodule
Perlit-Temperaturbereich/-Temperaturintervall pearlitic range
Perowskit perovskite
Perowskit-Strukturtyp perovskite [structure] type
Perowskit-Typ perovskite [structure] type
Pfad der bevorzugten Diffusion* short-circuit diffusion path
Phase phase
Phasenbestand phase composition
Phasenbestandteil phase constituent
Phasendiagram phase, equilibrium, or constitution diagram
Phasengesetz Gibbs' phase rule; phase rule
Phasengrenze interface; phase boundary
phasengrenzenbestimmte interface-controlled
Phasengrenzenenergie interfacial energy
Phasengrenzenversetzung misfit dislocation
Phasenkontrast phase contrast
Phasenregel phase rule
Phasenübergang phase transformation or transition
Phasenumwandlung phase transformation or transition
Phasenumwandlung 1.Ordnung first-order transition
Phasenumwandlung 2.Ordnung second-order transition

Photoelektronen-Spektroskopie electron spectroscopy for chemical analysis
  (ESCA)
physikalische Adsorption physical adsorption; physisorption
physikalische Eigenschaft physical property
physikalische Linienverbreiterung intrinsic [x-ray] line broadening
Piezoelektrikum piezoelectric
Pipe Diffusion pipe diffusion
plastische Verformung plastic deformation
Plattenmartensit acicular, plate, or lenticular martensite
Poisson-Zahl/-Konstante Poisson's ratio
Pol pole
Polarisationsfaktor polarization factor
Polarisationslichtmikroskopie polarized-light microscopy
Polfigur pole figure
Polnetz polar or equatorial net
polychromatische Röntgenstrahlung polychromatic [x-ray] or white radiation;
  Bremsstrahlung
Polygonisation polygonization
polygonisiert polygonized
polymorphe Kristallisation* polymorphic crystallization
polymorphe Modifikation polymorphic modification
polymorphe Umwandlung polymorphic transformation
Polymorphie polymorphism
Polytypie polytypism
Porosität porosity
Portevin-Le Chatelier-Effekt Portevin–Le Chatelier effect; dynamic strain
  aging
postdynamische Rekristallisation postdynamic recrystallization
Präzipitat precipitate
primäre Alpha-Phase [in Ti-Legierungen] primary α-phase [in Ti alloys]
primäres Gleitsystem active slip system
primäres Kriechen primary or transient creep
primäre Versetzung primary dislocation
Primärextinktion primary extinction
Primärgefüge primary structure
Primärkristallen primary crystals
Primärmischkristall primary or terminal solid solution
Primärrekristallisation primary recrystallization
Primärzwilling first-order twin
primitives Gitter primitive or simple lattice
prismatische Versetzungsschleife prismatic [dislocation] loop
Prismenebene prism plane
Prismengleitung prismatic slip
Probendickeneffekt* sample thickness effect
Pseudoplastizität pseudoplasticity
Pulverdiagram powder pattern

Pulverdiffraktometer diffractometer
Pulververfahren powder method
Punktdefekt point defect
Punktfehler point defect
punktförmiger Gitterfehler point defect
Punktgitter lattice; point or crystal lattice
Punktgruppe symmetry class; point group
Punktsymmetriegruppe symmetry class; point group
Pyramidenebene pyramidal plane
Pyramidengleitung pyramidal slip

Q

quantitative Metallographie quantitative metallography; stereology
Quarz quartz
quasiisotrop quasi-isotropic
Quasikristall quasicrystal
Quergleitung cross slip
Quer[kontraktions]zahl Poisson's ratio

R

Radialverteilungsfunktion* radial distribution function (RDF)
Raster-Durchstrahlungs-Elektronenmikroskop scanning transmission electron microscope (STEM)
Rasterelektronenmikroskop (REM) scanning electron microscope (SEM)
Rasterkraft-Mikroskop atomic force microscope (AFM)
Raster-Tunnel-Elektronenmikroskop (RTM) scanning tunneling microscope (STM)
Raumerfüllung packing factor
Raumgitter point or crystal lattice; lattice
Raumgruppe space group
raumzentriertes Gitter body-centered lattice
Reckalterung strain aging
Reflexionskugel reflection sphere
Reflexionskurve rocking curve
regellos-orientiert nonoriented
rekonstruktive Umwandlung reconstructive transformation
Rekristallisation recrystallization (ReX)
Rekristallisationsdiagram recrystallization diagram
Rekristallisationsglühen recrystallization annealing
Rekristallisationskeim recrystallization nucleus
Rekristallisationskeimbildung an Teilchen* particle-stimulated nucleation (PSN)
Rekristallisationstemperatur recrystallization temperature

Rekristallisationstextur recrystallization texture
Rekristallisationszwilling annealing twin
rekristallisiert recrystallized
Relaxation stress relaxation
Relaxationszeit relaxation time
relaxierter Modul relaxation modulus
Restaustenit residual or retained austenite
Rest-Beta-Phase retained β-phase
Restspannung residual stress
Restwiderstand residual electrical resistance
resultierende Schubspannung resolved shear stress
retrograde Soliduskurve retrograde solidus
Reversibilität reversibility
reversible Anlassversprödung reversible temper brittleness
reziproke Polfigur inverse pole figure
reziprokes Gitter reciprocal lattice
RHEED reflection high-energy electron diffraction (RHEED)
rhombisches System orthorhombic system
rhomboedrisches System rhombohedral or trigonal system
Richtung der leichtesten Magnetisierung easy magnetization direction
R-Lage R-orientation
Rocking-Kurve rocking curve
Rodrigues-Vektor Rodrigues vector
roh green
Röntgenabsorptionskante x-ray absorption edge
Röntgenabsorptionsspektrum x-ray absorption spectrum
Röntgenbeugungslinie x-ray diffraction line
Röntgenbeugungsmaximum x-ray diffraction line
Röntgendiffraktogram diffractogram
Röntgenemissionsspektrum x-ray emission spectrum
Röntgenfluoreszenz x-ray fluorescence
Röntgenlinienbreite x-ray line width
[Röntgen]linienintensität x-ray line intensity
Röntgen-Mikrobereichanalyse electron probe microanalysis (EPMA)
Röntgenmikroskopie x-ray microscopy or topography
Röntgen-Phasenanalyse x-ray structure analysis
Röntgen-Photoelektronenspektroskopie x-ray photoelectron spectroscopy
        (XPS)
Röntgenspektroskopie x-ray spectroscopy
Röntgenstrahlungsbeugung x-ray diffraction (XRD)
Röntgenstrahlungsstreuung x-ray scattering
Röntgentopographie x-ray topography or microscopy
Rückbildung reversion; precipitate reversion
rückhaltende Kraft pinning force
Rückstreuelektronen-Beugungsbild electron back-scattered pattern (EBSP)
rücktreibende Kraft drag force

r-Wert r-value; strain ratio
r̄-Wert r̄-value; Lankford coefficient
R-Zentrum R-center

S

Sachssches Faktor Sachs factor
Säulengefüge columnar structure
Säulenkristall columnar crystal or grain
Säulenzone columnar or transcrystallization zone
Scherband shear band
Schermodul modulus of rigidity; shear modulus
Schiebung shear strain
Schiebungsumwandlung shear or displacive transformation
Schmelzpunkt melting point or temperature (T_m)
Schmelzspinnen melt spinning; roller quenching
Schmelzspinn-Verfahren melt spinning; roller quenching
Schmelztemperatur melting point or temperature (T_m)
Schmelzverdüsung atomizing
Schmid-Faktor Schmid factor
Schmidsches Schubspannungsgesetz Schmid's law
Schottky-Paar Schottky pair
[Schräg] Beschattung shadowing
Schraubendisklination twist disclination
Schraubenversetzung screw dislocation
Schubbeanspruchung shear
Schubmodul shear modulus; modulus of rigidity
Schubspannung shear stress
Schwankung fluctuation
schwarzer Temperguß black-heart malleable [cast] iron
Schwarzkernguß black-heart malleable [cast] iron
Schwindung shrinkage
Segregat precipitated phase
Segregatbildung precipitation
Sehfeld field-of-view
Seigerung macrosegregation; major segregation; segregation; liquation
sekundäre Ausscheidung secondary precipitate
Sekundärelektron secondary electron
sekundäre Phase second phase
sekundäres Gleitsystem secondary slip system
sekundäres Kriechen steady-state or secondary creep
sekundäre Versetzung secondary dislocation
sekundäre Zeilengefüge banded structure
Sekundärextinktion secondary extinction
Sekundärgefüge secondary structure
Sekundärhärtung secondary hardening

Sekundärionen-Massenspektroskopie secondary ion mass spectroscopy
   (SIMS)
Sekundärkristall secondary crystal
Sekundärrekristallisation discontinuous, exaggerated, or abnormal grain
   growth; secondary recrystallization
Sekundärzementit proeutectoid or secondary cementite
Sekundärzwilling second-order twin
selbstähnlich self-similar
Selbstdiffusion self-diffusion
senkrechte Anisotropie normal anisotropy
sessile Versetzung sessile dislocation
Shockley-Teilversetzung Shockley partial dislocation
Sialon sialon
S-Ideallage S-orientation
Siliciumdioxid silica
Sintern sintering
skaleninvariant self-similar
S-Lage S-orientation
smektischer Kristall smectic crystal
S/Ms-Lage S/Bs-orientation
Snoek-Dämpfungsmaximum/-Relaxation Snoek peak or relaxation
Snoek–Köster-Dämpfungsmaximum/-Relaxation Köster peak or relaxation;
   Snoek–Köster peak or relaxation
Solidus[kurve] solidus
Solidus[linie] solidus
Solvus[kurve] solvus
Solvus[linie] solvus
Sonderkarbid special or alloy carbide
Sonderstahl alloy steel
Sorbit fine pearlite or sorbite
Sorption sorption
Spaltebene cleavage plane
Spannung stress
Spannungs-Dehnungs-Diagram stress-deformation diagram
Spannungsabbauglühen stress-relief anneal
Spannungsarmglühen stress-relief anneal
spannungsinduzierter Martensit stress-assisted or stress-induced martensite
spezielle Korngrenze special grain boundary; CSL-boundary
spezifische Grenzenfläche specific [interface] area
spezifisches Volumen specific volume
Sphalerit-Strukturtyp/-Typ sphalerite [structure] type
sphärische Abberation spherical aberration
Sphäroguß ductile [cast] iron
Sphärolyt spherulite
sphärolytscher Graphit nodular or spheroidal graphite
Spiegelebene mirror plane

Spinell spinel
Spinell-Strukturtyp spinel [structure] type
Spinell-Typ spinel [structure] type
Spinodale spinodal
spinodale Entmischung spinodal decomposition
spontan spontaneous
stabilisiertes Zirkoniumdioxid stabilized ZrO_2
Stahl steel
Stahl-Martensit steel martensite
Standarddreieck standard triangle; unit stereographic triangle
Standard-(*hkl*)-Projektion standard (*hkl*) projection
Stapelfehler stacking fault
Stapelfehlerenergie stacking-fault energy (SFE)
stationäres Kriechen secondary or steady-state creep
statische Gitterstörung static lattice distortion
statische Gitterverzerrung static lattice distortion
statische Erholung static recovery
statische Rekristallisation static recrystallization
Steinsalzgitter NaCl structure
Steinsalz-Strukturtyp/-Typ rock salt [structure] type
Stengelkristall columnar crystal
Stengelkristallzone columnar or transcrystallization zone
stereographische Projektion stereographic projection
stereographisches Netz stereographic net
Stereologie stereology; quantitative metallography
stetige Kornvergrößerung normal or continuous grain growth
stetiges Kornwachstum normal or continuous grain growth
Stöchiometrie stoiciometry or stoichiometric composition
strahleninduzierte Aushärtung irradiation hardening
strahleninduzierte Kristallbaufehler irradiation defects
strahleninduzierter Wachstum irradiation growth
strahleninduziertes Kriechen irradiation-induced creep
Strahlenschäden irradiation or radiation damage
Strained-Layer-Epitaxie strained-layer epitaxy
Stranski-Krastanov-Wachstum [modus] Stranski–Krastanov growth mode
Streckgrenze yield strength
Streifenkontrast wedge or thickness fringes
Struktur structure
Strukturamplitude structure factor or amplitude
strukturelle Fehlordnung structural disorder
strukturelle Leerstelle structural vacancy
Strukturfaktor structure factor or amplitude
Stückseigerung major or zonal segregation; macrosegregation; liquation
Stufenregel rule of stages
Stufenversetzung edge dislocation
Subgefüge substructure

Subkorn subgrain
Subkorngefüge subgrain structure
Subkorngrenze low-angle or subgrain boundary; subboundary
Subkornkoaleszenz subgrain coalescence
subkritisches Glühen subcritical annealing
submikrokristallin ultra-fine-grained
Substitutionsatom substitutional atom
Substitutionsmischkristall substitutional solid solution
Superplastizität superplasticity
Suzuki-Atmosphäre Suzuki atmosphere
Symmetrieachse symmetry axis
Symmetrieelement symmetry element
Symmetrieklasse symmetry class; point group
Symmetrieoperation symmetry operation
symmetrische Korngrenze symmetric boundary
System system

T

Tannenbaumkristall dendrite
Taylor-Faktor/-Orientierungsfaktor Taylor factor
technische Dehnung nominal strain
technische Spannung nominal stress
teilchenfreie Zone precipitation-free zone (PFZ)
Teilchenhärtung dispersion strengthening
Teilchenhemmung particle drag
Teilchen-Schneidemechanismus* particle shearing
Teilchenverfestigung dispersion strengthening
Teilchenvergröberung particle coarsening
teilgeordneter Mischkristall partially ordered solid solution
Teilgitter sublattice
teilkohärente Phasengrenze semicoherent or partially coherent interface
teilkohärentes Präzipitat, Teilchen partially coherent precipitate
Teilversetzung partial or imperfect dislocation
Temperatur vom Anfang der Baintbildung* (B_s-Temperatur) bainite start
  temperature (A_s)
Temperatur vom Anfang der Martensit-Austenit-Umwandlung* (A_s-Temperatur) austenite start temperature (A_s)
Temperatur vom Anfang der Martensitbildung* (M_s-Temperatur) martensite start temperature (M_s)
Temperatur vom Ende der Martensit-Austenit-Umwandlung* (A_f-Temperatur) austenite finish temperature (A_f)
Temperatur vom Ende der Martensitbildung* (M_f-Temperatur) martensite
  finish temperature (M_f)
Temperguß malleable [cast] iron
Temperkohle temper carbon

Tempern tempering [treatment]
tertiäre Rekristallisation tertiary recrystallization
tertiärer Zementit tertiary cementite
tertiäres Kriechen tertiary creep
Tetraedergitterlücke tetrahedral interstice, site, or void
Tetraederhohlraum tetrahedral interstice, site, or void
Tetraeder-Zwischengitterplatz tetrahedral interstice, site, or void
tetragonales System tetragonal system
Tetragonalität tetragonality
Tetrakaidekaeder tetrakaidecahedron
Textur preferred grain orientation; texture; crystallographic texture
Texturanalyse texture analysis
Texturbelegung texture intensity
texturiert grain-oriented; textured
Texturschärfe texture scatter; orientation spread
theoretische Festigkeit theoretical strength
thermisch-aktiviert thermally activated
thermische Analyse thermal analysis
thermische Ätzung vacuum or thermal etching
thermische Furche thermal groove
thermische Hysterese thermal hysteresis
thermisches Gräbchen thermal groove
thermische Stabilität thermal stability
thermodynamisches Gleichgewicht thermodynamic equilibrium
thermodynamische Stabilität thermodynamic stability
thermoelastischer Martensit thermoelastic martensite
thermoelastische Spannungen thermal stresses
thermomagnetische Behandlung thermo-magnetic treatment
thermomechanische Behandlung thermo-mechanical processing, treatment
Thompson-Tetraeder Thompson's tetrahedron
Thomson-Freundlich-Gleichung Thomson–Freundlich equation
Tiefenschärfe depth of focus
tiefe Störungsstelle deep center
tiefes Zentrum deep center
tief-temperatur-thermomechanische Behandlung low-temperature thermo-
        mechanical treatment; ausforming
Titan-Martensit titanium martensite
Transitionsband transition deformation band or microband
transkristallin transcrystalline; transgranular; intracrystalline; intragranular
Transkristallisationszone transcrystallization or columnar zone
Translationsgruppe translation group
Translationsperiode translation vector
Translationsvektor translation or base vector
Transmissions-Elektronenmikroskop (TEM) transmission electron micro-
        scope (TEM)
treibende Kraft driving force

Tridymit tridymite
Triebkraft driving force
trigonales System trigonal or rhombohedral system
triklines System triclinic system
Tripelpunkt triple junction; triple point
Troostit trostite
Tweedgefüge modulated structure

U

Überalterung overaging
übereutektisch hypereutectic
übereutektoid hypereutectoid
Übergangskriechen transient or primary creep
Überhitzung superheating
Übersättigung supersaturation
Überstruktur superlattice; ordered solid solution; superstructure
Überstrukturversetzung superdislocation
ultrafeinkörnig ultra fine-grained
umgewandelte Beta-Gefüge transformed β structure (β_{tr})
Umklapptransformation displacive transformation
Umklappumwandlung shear[-type] transformation
Umlagerungsbereich displacement cascade
Umlösung Ostwald ripening
Umwandlungsgeschwindigkeit transformation rate
Umwandlungshysterese transformation hysteresis
umwandlungsinduzierte Plastizität transformation-induced plasticity (TRIP)
Umwandlungskinetik kinetics [of transformation]
Umwandlungspunkt critical point
Umwandlungsspannungen transformation stresses
Umwandlungstemperatur arrest point
Umwandlungstemperaturbereich transformation range
Umwandlungsverstärkung transformation toughening
Umwandlungszwilling transformation twin
unelastische Streuung inelastic scattering
ungeordneter Mischkristall disordered or random solid solution
unlegierter Stahl plain carbon steel
unstetige Kornvergrößerung exaggerated, discontinuous, or abnormal grain
    growth; secondary recrystallization
unstetiges Kornwachstum exaggerated, discontinuous, or abnormal grain
    growth; secondary recrystallization
unsymmetrische Korngrenze asymmetric boundary
unterer Bainit lower bainite
untere Streckgrenze lower yield stress
untereutektisch hypoeutectic
untereutektoid hypoeutectoid

untereutektoider Ferrit proeutectoid ferrite
Untergitter sublattice
Untergrund background
unterkühlt undercooled
Unterkühlung supercooling; undercooling
unumkehrbar irreversible
unvollständige Versetzung partial or imperfect dislocation

V

Vakuumätzung vacuum or thermal etching
Valenzband valence band
Valenzelektronenkonzentration electron:atom ratio; electron concentration
Van-der-Waalssche Bindung van der Waals bond
Vegardsche Regel Vegard's law
Verankerungskraft pinning force
verbotene Band forbidden gap
Verbundwerkstoff composite
verdichteter Graphit vermicular or compacted graphite
verdünnte [feste] Lösung dilute [solid] solution
Veredelung modification
Vereinigung coalescence
Verfälschung artifact
Verfestigung work hardening
Verfestigungsexponente strain-hardening exponent
Verformung strain
Verformungsalterung strain aging
Verformungsgeschwindigkeit strain rate
Verformungsgrad strain
verformungsinduzierte Korngrenzenwanderung* strain-induced grain
    boundary migration (SIBM)
Verformungsknicken deformation kinking
Verformungsmartensit strain-induced martensite
Verformungsmechanismus deformation mechanism
Verformungsmechanismus-Schaubild deformation mechanism map
Verformungstextur deformation texture
Verformungsverfestigung strain hardening; work hardening
Verformungszwilling deformation twin
Vergröberung coarsening
Vergrößerung magnification
vergütbar heat treatable
Verlagerungskaskade displacement cascade
vermikularer Graphit compacted or vermicular graphite
Versetzung dislocation
Versetzungsaufbau dislocation structure; fine structure
Versetzungsauflöschung dislocation annihilation

Versetzungsaufspaltung dislocation dissociation or splitting
Versetzungsaufstau pile-up
Versetzungsausheilung dislocation annihilation
Versetzungsbreite dislocation width
Versetzungsdelokalisation dislocation delocalization
Versetzungsdichte dislocation density
Versetzungsdickicht dislocation tangle
Versetzungsdipole dislocation dipole
Versetzungsenergie dislocation energy
Versetzungskern dislocation core
Versetzungskinke kink
Versetzungsklettern climb
Versetzungskriechen dislocation creep
Versetzungslinienspannung dislocation line tension
Versetzungsnetz dislocation network
Versetzungsquelle dislocation source
Versetzungsschleife dislocation loop
Versetzungsspannungsfeld dislocation stress field
Versetzungssprung jog
Versetzungsverankerung dislocation pinning
Versetzungsvermehrung dislocation multiplication
Versetzungsvervielfachung dislocation multiplication
Versetzungsvorzeichen dislocation sense
Versetzungswand dislocation wall
Versetzungswendel helical dislocation
Verunreinigung impurity
Verunreinigungshemmung* impurity drag
Vielkristall polycrystal
Vielkristalldiffraktometer diffractometer
vielkristallin polycrystalline
Viskoelasizität viscoelasticity
vizinale Oberfläche vicinal plane
Vollmer–Weber-Wachstum[modus] Vollmer–Weber growth mode
vollständige Versetzung perfect dislocation
Vollständigglühen full annealing
Volumenanteil; Vol-% vol%
Volumendiffusion bulk, lattice, or volume diffusion
Volumenkontraktion shrinkage
Vor-Ausscheidung pre-precipitation
voreutektoider proeutectoid
voreutektoider Ferrit proeutectoid ferrite
voreutektoider Zementit secondary or proeutectoid cementite
vorgebildeter Keim preformed nucleus
Vorkeim embryo
Vorlegierung master alloy; alloy composition
Vorzugslage texture component

Vorzugsorientierung texture; preferred grain orientation; crystallographic texture

W

Wagner-Lifshitz-Slyozov-Theorie Wagner–Lifshitz–Slyozov theory
wahre Dehnung, Verformung true strain
wahre Spannung true stress
wahre Spannungs-wahre Dehnungs-Kurve stress-strain diagram
Waldversetzung forest dislocation
Warmauslagern artificial aging
Wärmebehandlung heat or thermal treatment
warmverformt warm or hot worked
Warmverformung warm or hot deformation
Warren-Averbach-Verfahren Warren–Averbach method
Weak-Beam-Dunkelfeldabbildung weak-beam imaging
Weg der bevorzugten Diffusion* short-circuit diffusion path
weißer Temperguß white-heart malleable [cast] iron
weißes Gußeisen white [cast] iron
Weichglühen annealing or anneal; softening or full annealing
Weisscher Zonengesetz Weiss zone law
Weissches Bezirk/Elementarbereich magnetic domain
weitreichend long-range
wellenlängendispersive Spektrometrie wavelength-dispersive spectrometry (WDS)
Wellenlängenspektrum wavelength spectrum
Widmannstättenscher Ferrit Widmannstätten ferrite
Widmannstättensches Gefüge Widmannstätten structure
Wirtsatom host atom
Wirtsstoff solvent
Wulff-Plot σ-plot
Wulffsches Netz Wulff net
Würfellage cube orientation
Würfeltextur cube texture
Wurzit wurzite

Y

Youngsches Modul Young's modulus

Z

Zeilenkarbide carbide stringers
Zeit-Temperatur-Umwandlungs-/ZTU-Schaubild time-temperature-transformation (TTT) diagram

Zellenausscheidung discontinuous precipitation
Zellengefüge striation or cellular structure
Zellensubgefüge lineage structure
Zellgefüge cell structure
zellulare Mikroseigerung cellular microsegregation
Zementit cementite ·
Zener-Dämpfungsmaximum/-Relaxation Zener peak or relaxation
Zenersche Hemmungskraft Zener drag
Zerstäubung atomizing or sputtering
Zinkblendegitter ZnS cubic structure
Zinkblende-Strukturtyp/-Typ zincblende [structure] type
Zipfelbildung earing zirkoniumoxid-verstärktes Aluminiumoxid zirconia-
    toughened alumina (ZTA)
Zone zone
Zonenachse zone axis
Zonenglühen zone annealing
ZTU-Diagram time-temperature-transformation (TTT) diagram
ZTU-Schaubild für kontinuierliche Abkühlung continuous cooling transfor-
    mation (CCT) diagram
Zugbeanspruchung tension
Zugspannung tensile stress
Zugumformung tension
Zusammensetzung composition
Zusatzelement alloying element
Zustandsdiagram equilibrium, constitution, or phase diagram
Zwang constraint
zweiphasiges Gefüge dual-phase microstructure
zweistoff binary
zweistufiger Abdruck indirect replica
Zweiweg-Formgedächtniseffekt two-way shape memory effect
Zwilling twin
Zwillingsbildungssystem twinning system
Zwillingskristall twinned crystal
Zwischengitterdiffusion interstitial [mechanism of] diffusion
Zwischengitterplatz interstice; lattice void
Zwischenphase intermediate; transition; or transient phase
Zwischenstufengefüge bainite
Zwischenstufenumwandlung bainitic transformation
Zwischenstufenvergütung austempering

Literature For Further Reading

Ashby, M. F. and Jones, D. R. H., *Engineering Materials*, 1st and 2nd pts, Butterworth-Heinemann, Oxford, 1996, 1998.

Barrett, C. and Massalski, T. B., *Structure of Materials*, Pergamon, Oxford, 1980.

Briggs, D. and Seah, M. P., Eds., *Practical Surface Analysis*, v.1 and 2, Wiley, New York, 1990, 1992.

Cahn, R. W. and Haasen, P., Eds., *Physical Metallurgy*, North Holland, 1996.

Chiang, Y.-M., Birnie III, D. P., and Kingery, W. D. *Physical Ceramics*, John Wiley & Sons, New York, 1997.

Christian, J. W., *The Theory of Transformation in Metals and Alloys*, Pergamon, Oxford, 1975.

Collings, E. W., *The Physical Metallurgy of Titanium Alloys*, ASM, Metals Park, 1984.

Cullity, B. D., *Elements of X-Ray Diffraction*, Addison-Wesley, Reading, MA, 1978.

Dowben, P. A. and Miller, A., Eds., *Surface Segregation Phenomena*, CRC, Boca Raton, FL, 1990.

Diwling, N. E., *Mechanical Behaviour of Materials*, Prentice Hall, New York, 1999.

Duke, P. J. and Michette, A. G., Eds., *Modern Microscopies*, Plenum Press, New York, 1990.

Feldman, L. C., Mayer, J. W., and Picraux, S. T., *Materials Analysis by Ion Channeling*, Academic Press, San Diego, 1982.

Goldstein, J. L., et al. *Scanning Electron Microscopy and X-Ray Microanalysis*, Plenum Press, New York, 1994.

Hamphreys, F. J. and Hatherly, M., *Recrystallization and Related Annealing Phenomena*, Pergamon, Oxford, 1995.

Heimendahl, M., *Einführung in Elektronenmikroskopie*, Vieweg, Braunschweig, 1970.

Honeycomb, R. W. K. and Bhadeshia, H. K. D. H., *Steels*, Arnold, London, 1995.

Hull, D. and Bacon, D. J., *Introduction to Dislocations*, Pergamon, Oxford, 1989.

Klug, H. P. and Alexander, L. E., *X-Ray Diffraction Procedures*, Wiley & Sons, New York, 1974.

Kurzydlowski, K. J. and Ralph, B., *Quantitative Description of Microstructure of Materials*, CRC, Boca Raton, FL, 1995.

Liebermann, H. H., Ed., *Rapidly Solidified Alloys*, Wiley & Sons, New York, 1993.

Lifshin, E., Ed., *Characterization of Materials*, 1st and 2nd pts, VCH, Weinheim, 1992, 1994.

Martin J. W., *Precipitation Hardening*, Pergamon, Oxford, 1968.

Martin, J. W., Cantor, B., and Doherty, R. D., *Stability of Microstructure in Metallic Systems*, University Press, Cambridge, 1996.

Meyers, M. A. and Chawla, K. K., *Mechanical Behavior of Materials*, Prentice Hall, New Jersey, 1999.

Murr, L. E., *Interfacial Phenomena in Metals and Alloys*, Addison-Wesley, Reading, MA, 1975.

Newbury, D. E., et al., *Advanced Scanning Electron Microscopy and X-Ray Microanalysis*. Plenum Press, New York, 1986.

Novikov, V., *Grain Growth and Control of Microstructure and Texture in Polycrystalline Materials*, CRC, Boca Raton, FL, 1996.

Polmear, I. L., *Light Alloys*, Arnold, London, 1995.

Reed-Hill, R. E. and Abbaschian, R., *Physical Metallurgy Principles*, PWS-Kent, Boston, 1992.

Rhines, F. N., *Microstructurology: Behaviour and Microstructure of Materials*, Riederer, Stuttgart, 1986.

Russel, K. C. and Aaronson, H. I., Eds., *Precipitation Processes in Solids*, Met. Soc. AIME, Warrendale, PA, 1978.

Smallman, R. E. and Bishop, R. J., *Modern Physical Metallurgy and Materials Engineering*, Butterworth-Heinemann, Oxford, 1999.

Sutton, A. and Baluffi, R. W., *Interfaces in Crystalline Materials*, University Press, Oxford, 1995.

Svensson, L.-E., *Control of Microstructures and Properties in Steel Arc Welds*, CRC, Boca Raton, FL, 1994.

Thomas, G. and Goringe, M. J., *Transmission Electron Microscopy of Materials*, Wiley & Sons, New York, 1979.

Underwood, E. E., *Quantitative Stereology*, Addison-Wesley, Reading, MA, 1970.

Walls, J. M., Ed., *Methods of Surface Analysis*, University Press, Cambridge, 1992.

Williams, D. B., Pelton, A. R., and Gronsky, R., *Images of Materials*, University Press, Oxford, 1991.

Wojnar, L., *Image Analysis*, CRC, Boca Raton, FL, 1999.

Literature For Further Reading

Ashby, M. F. and Jones, D. R. H., *Engineering Materials*, 1st and 2nd pts, Butterworth-Heinemann, Oxford, 1996, 1998.

Barrett, C. and Massalski, T. B., *Structure of Materials*, Pergamon, Oxford, 1980.

Briggs, D. and Seah, M. P., Eds., *Practical Surface Analysis*, v.1 and 2, Wiley, New York, 1990, 1992.

Cahn, R. W. and Haasen, P., Eds., *Physical Metallurgy*, North Holland, 1996.

Chiang, Y.-M., Birnie III, D. P., and Kingery, W. D. *Physical Ceramics*, John Wiley & Sons, New York, 1997.

Christian, J. W., *The Theory of Transformation in Metals and Alloys*, Pergamon, Oxford, 1975.

Collings, E. W., *The Physical Metallurgy of Titanium Alloys*, ASM, Metals Park, 1984.

Cullity, B. D., *Elements of X-Ray Diffraction*, Addison-Wesley, Reading, MA, 1978.

Dowben, P. A. and Miller, A., Eds., *Surface Segregation Phenomena*, CRC, Boca Raton, FL, 1990.

Diwling, N. E., *Mechanical Behaviour of Materials*, Prentice Hall, New York, 1999.

Duke, P. J. and Michette, A. G., Eds., *Modern Microscopies*, Plenum Press, New York, 1990.

Feldman, L. C., Mayer, J. W., and Picraux, S. T., *Materials Analysis by Ion Channeling*, Academic Press, San Diego, 1982.

Goldstein, J. L., et al. *Scanning Electron Microscopy and X-Ray Microanalysis*, Plenum Press, New York, 1994.

Hamphreys, F. J. and Hatherly, M., *Recrystallization and Related Annealing Phenomena*, Pergamon, Oxford, 1995.

Heimendahl, M., *Einführung in Elektronenmikroskopie*, Vieweg, Braunschweig, 1970.

Honeycomb, R. W. K. and Bhadeshia, H. K. D. H., *Steels*, Arnold, London, 1995.

Hull, D. and Bacon, D. J., *Introduction to Dislocations*, Pergamon, Oxford, 1989.

Klug, H. P. and Alexander, L. E., *X-Ray Diffraction Procedures*, Wiley & Sons, New York, 1974.

Kurzydlowski, K. J. and Ralph, B., *Quantitative Description of Microstructure of Materials*, CRC, Boca Raton, FL, 1995.

Liebermann, H. H., Ed., *Rapidly Solidified Alloys*, Wiley & Sons, New York, 1993.

Lifshin, E., Ed., *Characterization of Materials*, 1st and 2nd pts, VCH, Weinheim, 1992, 1994.

Martin J. W., *Precipitation Hardening*, Pergamon, Oxford, 1968.

Martin, J. W., Cantor, B., and Doherty, R. D., *Stability of Microstructure in Metallic Systems*, University Press, Cambridge, 1996.

Meyers, M. A. and Chawla, K. K., *Mechanical Behavior of Materials*, Prentice Hall, New Jersey, 1999.

Murr, L. E., *Interfacial Phenomena in Metals and Alloys*, Addison-Wesley, Reading, MA, 1975.

Newbury, D. E., et al., *Advanced Scanning Electron Microscopy and X-Ray Microanalysis*. Plenum Press, New York, 1986.

Novikov, V., *Grain Growth and Control of Microstructure and Texture in Polycrystalline Materials*, CRC, Boca Raton, FL, 1996.

Polmear, I. L., *Light Alloys*, Arnold, London, 1995.

Reed-Hill, R. E. and Abbaschian, R., *Physical Metallurgy Principles*, PWS-Kent, Boston, 1992.

Rhines, F. N., *Microstructurology: Behaviour and Microstructure of Materials*, Riederer, Stuttgart, 1986.

Russel, K. C. and Aaronson, H. I., Eds., *Precipitation Processes in Solids*, Met. Soc. AIME, Warrendale, PA, 1978.

Smallman, R. E. and Bishop, R. J., *Modern Physical Metallurgy and Materials Engineering*, Butterworth-Heinemann, Oxford, 1999.

Sutton, A. and Baluffi, R. W., *Interfaces in Crystalline Materials*, University Press, Oxford, 1995.

Svensson, L.-E., *Control of Microstructures and Properties in Steel Arc Welds*, CRC, Boca Raton, FL, 1994.

Thomas, G. and Goringe, M. J., *Transmission Electron Microscopy of Materials*, Wiley & Sons, New York, 1979.

Underwood, E. E., *Quantitative Stereology*, Addison-Wesley, Reading, MA, 1970.

Walls, J. M., Ed., *Methods of Surface Analysis*, University Press, Cambridge, 1992.

Williams, D. B., Pelton, A. R., and Gronsky, R., *Images of Materials*, University Press, Oxford, 1991.

Wojnar, L., *Image Analysis*, CRC, Boca Raton, FL, 1999.